CHEMICAL INTERACTIONS
RESOURCES

IMAGES, DATA, AND READINGS

DEVELOPED AT LAWRENCE HALL OF SCIENCE, UNIVERSITY OF CALIFORNIA AT BERKELEY
PUBLISHED AND DISTRIBUTED BY DELTA EDUCATION

FOSS Middle School Staff and Associates

FOSS Middle School Curriculum Development Team

Linda De Lucchi and Larry Malone, Co-Directors
Teri Dannenberg, Curriculum Developer; Ann Moriarty, Curriculum Developer
Dr. Susan Ketchner, Content Specialist; Susan Kaschner Jagoda, Curriculum Developer
Virginia Reid, Curriculum Consultant; Dr. Terry Shaw, Curriculum Consultant
Dr. Kathy Long, Assessment Coordinator
Carol Sevilla, Publications Coordinator; Rose Craig, Illustrator
Alev Burton, Program Assistant

ScienceVIEW Multimedia Design Team

Alana Chan, Project Manager; Leigh Anne McConnaughey, Principal Illustrator
Dan Bluestein, Lead Programmer; Chris Cianciarulo, Programmer
Roger Vang, Programmer; Nicole Medina, Production Assistant

Special Contributors

Marshall Montgomery, Materials Design; John Quick, Photography and Video
Barbara Stebbins, 7th-grade teacher, Black Pine Circle School, Berkeley, CA, and her students
Advisers: Jennifer Claesgens, John Sheridan, Dr. Angelica Stacy, Lillie Tucker Akin

Delta Education FOSS Middle School Team

Bonnie Piotrowski, FOSS Editorial Director; Mathew Bacon, Vice President, Product Development
Project Team: Megan Dumm, Tom Guetling, Joann Hoy, Haley January, Lisa Lachance,
Angela Miccinello, Cathrine Monson, John Prescott, Barbara Resch, Jennifer Roehrig, Nina Whitney

National Trial Teachers

Seth Corrigan, M.L. King Middle School, Berkeley, CA
Michael Romero, Angerine Middle School, Boulder, CO
Stephen "Doc" Holaday, Abilene MS, Abilene, KS
Jessica Penchos, Timilty MS, Roxbury, MA
Cheryl Hall, Grover Cleveland MS, Dorchester, MA
Sarah Chapin and Joanna Snyder, Hudson High School, Hudson, MA
Gayle Dunlap, Saddle Brook Middle School, Saddle Brook, NJ
Corinne De Keukela, Walter T. Bergen MS, Bloomingdale, NJ
Andrew Kramar, Cavallini Middle School, Upper Saddle River, NJ
Beata Ferris, Belle Fourche School District, Belle Fourche, SD
Steven Barry, Spring Branch MS, Houston, TX
Charlene D. Guerra, Landrum MS, Houston, TX
Diane Mayer and Jason A. Schmid, Oak Harbor MS, Oak Harbor, WA
Jack Aldridge, Thurgood Marshall MS, Olympia, WA
Melissa Barnard and Anna Horton, Nelsen MS, Renton, WA
Lori Aegerter and Sheryl Dhuey, Cherokee Heights MS, Madison, WI
Suzanne M. Folberg and Mary E. Joe, O'Keeffe MS, Madison, WI

FOSS for Middle School Project
Lawrence Hall of Science, University of California
Berkeley, CA 94720   510-642-8941

...because children learn by doing.®

Delta Education, LLC
P.O. Box 3000 80 Northwest Blvd.
Nashua, NH 03063
1-800-258-1302

The FOSS Middle School Program was developed in part with the support of the National Science Foundation Grant ESI 9553600. However, any opinions, findings, conclusions, statements, and recommendations expressed herein are those of the authors and do not necessarily reflect the views of the NSF.

Chemical Interactions

542-1505

Printing 8 — 12/2012
Quad/Graphics, Versailles, KY

ISBN-10: 1-58356-444-6
ISBN-13: 978-1-58356-444-8

FOSS Chemical Interactions Resources

Table of Contents

Readings

References

ELEMENTS

Two thousand years ago, people were trying to figure out what things were made of. One idea was that everything was a mix of four basic properties: hot, cold, wet, and dry.

If you had just the right mix of hot and dry, that might make rock. A little less hot and a bit of wet might make wood. The right amount of all four properties might make a leaf.

Pure samples of the four properties were fire, air, earth, and water. These four substances were thought to be the elements from which everything was made. A table of the ancient elements looked like this.

Some people had a different idea about what things were made of. Chemists in the 1800s were busy investigating a lot of different **substances**. They heated substances as hot as they could. They put acid on them. They ran electric currents through them. Sometimes the substances separated into new substances when they did their experiments. When this happened, they tested the new substances with heat, acid, and electricity. Some of the substances would not change any more. They called the unchangeable substances **elements**. These new elements had different names than the ancient elements. The new elements had names like iron, copper, carbon, oxygen, sulfur, and gold.

An element is a **fundamental** substance that cannot be broken into simpler substances. Elements are the building blocks of **matter**. Elements combine to form all the different substances in the world.

By the middle of the 1800s, about 60 elements had been discovered. A lot was known about them. Scientists knew some of their **chemical properties**, such as what other elements they combine with. They knew some of their **physical properties**, such as the weight of standard samples of the elements. When scientists made a list of the elements, they put them in order by weight, starting with the lightest element they knew about, hydrogen.

THE FIRST PERIODIC TABLE

In 1869, a Russian chemist named Dmitry Ivanovich Mendeleyev (1834–1907) was writing a book about the elements. He made a set of element cards. Each card had one element's name and symbol and everything that was known about it. He put the cards in one long row from lightest to heaviest, hydrogen to uranium.

Mendeleyev looked at the line of element cards and saw something interesting. The first two elements, hydrogen (H) and lithium (Li), had similar chemical properties.

And as he looked down the line, he noticed that sodium (Na) and potassium (K) also had chemical properties similar to hydrogen and lithium. The similar chemical properties showed up periodically in his lineup.

Then Mendeleyev saw that beryllium (Be), magnesium (Mg), and calcium (Ca) all had similar, but different properties. The similar chemical properties of beryllium, magnesium, and calcium showed up periodically, too.

Mendeleyev had an idea. He reorganized the cards into several short rows. This way all the elements with similar properties lined up in columns. The columns are called groups.

The periodic recurrence of similar chemical properties is why the element table is called the periodic table of the elements.

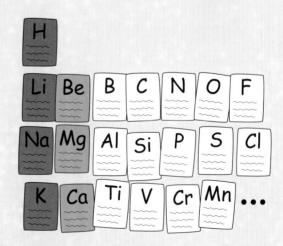

Dmitry Ivanovich Mendeleyev

When Mendeleyev had all the elements laid out, he noticed something was wrong. For instance, the chemical properties of titanium (Ti) were not like those of aluminum (Al) and boron (B) above it.

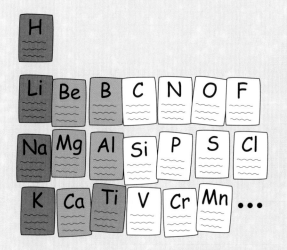

When Mendeleyev moved titanium and its neighbors to the right, two things happened. The chemical properties of the elements lined up better. And there was a gap in the table of elements.

Mendeleyev looked at the gap and **predicted** that an undiscovered element must fit in that spot. Furthermore, he predicted the properties that the new element would have. By moving the known elements around so that their properties lined up, Mendeleyev predicted about 30 new elements. Over the next 30 years, most of them were discovered.

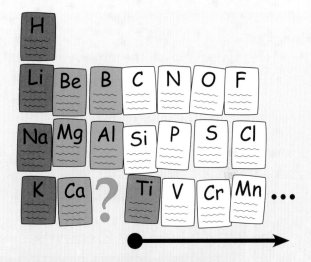

THE MODERN PERIODIC TABLE OF THE ELEMENTS

The modern **periodic table of the elements** organizes and displays all the elements from simplest to most complex. Hydrogen, the simplest element, is number 1. Mendeleyev's idea of putting the elements in rows under each other, so that the chemical properties are similar in the columns, is still used. But Mendeleyev didn't know what we know today. There are 2 elements in row 1, 8 elements in rows 2 and 3, 18 elements in rows 4 and 5, and 32 elements in rows 6 and 7. This is the modern periodic table.

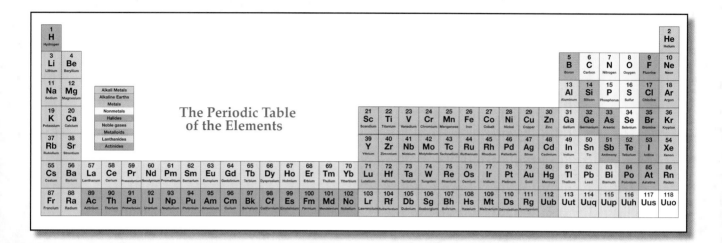

This layout makes the table very long. Often 28 of the elements are pulled out and shown below the others. Then the table fits better on a standard piece of paper.

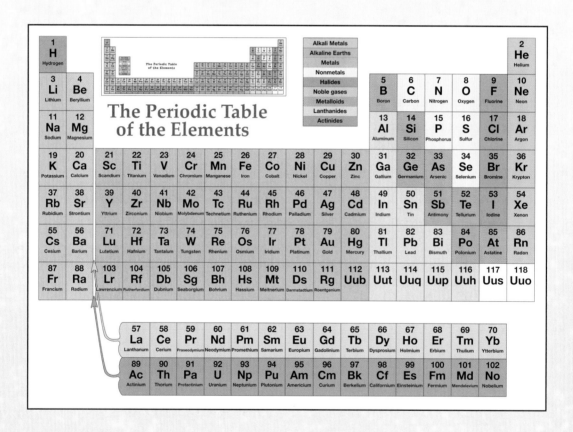

Color helps to show which elements have similar chemical properties. In the periodic table used in this course, elements that are orange, light blue, dark blue, light lavender, dark lavender, and yellow green are **metals**. Green, red, yellow, and aqua elements (plus hydrogen) are nonmetals. The green elements on the right are called **noble gases**. They are interesting because they don't react with other elements.

6

ELEMENT FINDERS

Sir Humphry Davy (1778–1829)

Sir Humphry Davy was born in Cornwall, England, in 1778. As a young man, he was studying to be a doctor. But his life changed when he picked up a book on chemistry.

Sir Humphry Davy

Soon Davy was conducting experiments in his small laboratory. Some of his first efforts ended in explosions. Others filled his lab with strange gases. Davy's knowledge of chemistry grew. He took a teaching position at the university when he was 24 years old. There he became interested in separating substances until they could not be separated any more. He used a battery to run electricity through a solution of **potash**, which is potassium carbonate, K_2CO_3. The potash separated.

Davy discovered his first element, potassium (K). People say that Davy actually danced around the room after this discovery.

Davy went on to become one of the greatest element finders of all time. He is credited with more element discoveries than anyone else! Using his electricity methods, Davy discovered seven elements, including some you've read about. They are sodium (Na), magnesium (Mg), boron (B), potassium (K), calcium (Ca), barium (Ba), and chlorine (Cl).

Marie Curie (1867–1934)

Marie Sklodowska Curie was born in Warsaw, Poland, on November 7, 1867. In 1891, she moved to Paris, France, to study mathematics, physics, and chemistry. After getting her degree in physics, she set up a small lab in the basement of the school where her husband, Pierre, taught. She studied the **radiation** coming from uranium ore. Curie thought that the amount of radiation was too strong to be only from uranium. She discovered two new elements, polonium (Po) and radium (Ra), in the ore sample. The samples of radium she produced glowed with a continuous green light. She invented the term *radioactivity* to describe the radiation given off by the elements.

Marie Curie

REVIEW QUESTIONS

1. What is an element?

2. How are matter and elements related?

3. How was Mendeleyev able to predict the existence of elements that had not yet been discovered?

4. What is the periodic table of the elements?

Curie was awarded the Nobel Prize in 1903. She was the first woman ever to win. She was awarded a second Nobel Prize in 1911, making her the first person ever to win twice! During World War I (1914–1918), Curie trained people to use X rays to find bullets in wounded soldiers.

Unfortunately, Curie didn't realize the dangers of radiation. In 1934, she died from an illness caused by exposure to radioactive materials. Her notes and laboratory equipment are still radioactive today, more than 100 years after she conducted her research.

Elements in the Universe

About 5 billion years ago, a huge star exploded. The blast sent a giant cloud of gas and dust into space. Over the next few million years, gravity pulled the bits of gas and dust closer together. Finally, the tiny particles formed the Solar System. The Sun, the planets, and everything on them came from that space cloud. Everything in the world, including you, is made of stardust.

What was that stardust made of? Elements. Particles of all 90 naturally occurring elements were flying around in that space cloud. And when Earth formed, all 90 elements became part of our planet.

Elements in the Sun

Our star, the Sun, has **mass** and occupies space. It is matter. All matter is made of elements. The Sun is no exception.

If you could separate the Sun's elements and put them in piles, the pile of hydrogen would be the largest. The Sun is about 75% hydrogen by weight. The next largest pile would be helium. The Sun is about 23% helium. The next three elements are in very small piles: 0.9% oxygen, 0.3% carbon, and 0.1% nitrogen. The Sun is mostly hydrogen and helium with small amounts of other elements.

elements in the periodic table that you never heard of are the rare ones.

The most abundant Earth elements

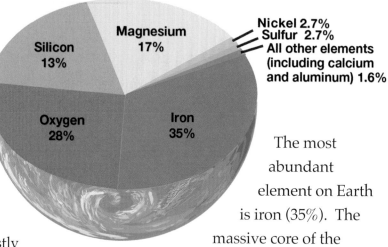

The most abundant element on Earth is iron (35%). The massive core of the planet is mostly iron. Next are oxygen (28%), magnesium (17%), and silicon (13%). These three elements are the main elements in minerals and rocks. They make up the largest part of the planet, the **mantle** and **crust**. The other major elements that make up Earth are nickel (2.7%), sulfur (2.7%), calcium (0.6%), and aluminum (0.4%). The remaining 82 elements together make up a tiny 0.6% of Earth.

It might seem that Earth is a pretty simple place. It's made mostly out of a dozen or so elements. But Earth is not a simple place. It is not the number of elements that determines how complex things are. It's the ways the elements combine to make different substances. These few elements can combine to make millions of different materials. That's where the wonderful variety on Earth comes from.

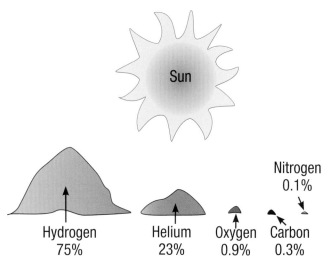

The most abundant elements in the Sun

Earth Elements

The planet we live on, Earth, is made of all 90 naturally occurring elements. But the elements are not all found in equal abundance. Some elements are common, and others are extremely rare. Those

Elements in the Sky

The **atmosphere** surrounding Earth is matter in its gas phase. Matter is made of elements. What elements are in the atmosphere?

The atmosphere is mostly **nitrogen** (78%). The second-most abundant element is oxygen (21%). The third-most common element is the noble gas argon. Less than 1% of the atmosphere is argon.

The other elements in the air are present in small quantities. **Carbon dioxide gas** in the air has the element carbon. Water in the air has the element hydrogen. Smoke from fires and exhaust from industries and motor vehicles add more elements to the air. Some of these elements can damage plants and animals. Mercury, lead, and some substances containing sulfur and nitrogen can be health hazards.

Elements in the Ocean

The ocean covers almost three-quarters of Earth's surface. And the ocean is deep. That adds up to a lot of water, and water is made of two elements. For this reason, the ocean is 85% oxygen and 11% hydrogen. Because seawater is salty, it contains a lot of other elements. The most common **salt** is sodium chloride (NaCl). Many other kinds of salt are dissolved in the sea, too. As a result, chlorine (2%) is the third-most abundant element in the sea, and sodium (1%) is the fourth-most abundant.

All 86 of the other elements are in the sea as well. For billions of years, water has been washing across the land, flowing into rivers, and finally making its way to the sea. All of the elements dissolved by water on land end up in Earth's ocean. That means there is gold in seawater. The problem is how to get it out.

Earth's ocean and sky

Elements in You

You are made of elements in the periodic table. How many of the 90 elements do you think it takes to make a person?

You probably have a trace amount of every element in your body. That's because elements are found everywhere, including our air, water, and food. For instance, helium is in the air in tiny amounts. Small amounts of helium enter our bodies when we breathe. We don't need helium to survive, but it's there.

Other elements are essential for life. We need tiny amounts of some elements, like chlorine and iodine. But we need large amounts of others, such as carbon and oxygen.

The human body is about 75% water. Water is made from the elements hydrogen and oxygen.

Much of the solid mass of your body is made from the element carbon combined with other elements. Carbon, hydrogen, and oxygen combine to form **carbohydrates** (sugars and starches), **lipids** (oils and fats), and **proteins**. Proteins also contain nitrogen. Skin, muscle, fat, and organs are made of carbohydrates, lipids, and proteins.

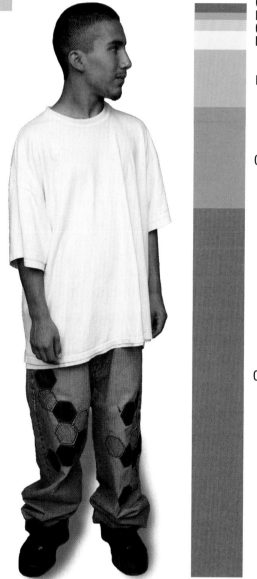

Other 1.5%
Phosphorus 1.0%
Calcium 1.5%
Nitrogen 3.0%

Hydrogen 10%

Carbon 18%

Oxygen 65%

The tough, rigid parts of the body, like teeth, bones, and cartilage, are rich in the element calcium. Blood contains a lot of iron. Potassium and sodium are needed for nerve and brain function.

When you add it all up, about 98.5% of the human body is composed of only six elements. They are oxygen, carbon, hydrogen, nitrogen, calcium, and phosphorus. The remaining 1.5% is small amounts of a lot of different elements.

Most living organisms are chemically very similar to humans. So it's pretty safe to say that all life is based on the same six elements. The trees and other plants are mostly carbon, hydrogen, and oxygen. So are the birds, insects, and spiders living in and among the plants. And anything we use that comes from organisms, such as wood, leather, paper, cotton, fabrics, plastics, food, and fuels, is also made of these six elements.

So what elements are found in the universe? The answer is all of them. And which elements are abundant? Only a few. Common matter is made mostly of just a handful of elements.

Review Questions

1. What element is among the five most abundant elements in the Sun, Earth, ocean, atmosphere, and organisms?

2. What does it mean when people say everything is made of stardust?

3. Why are the elements carbon, hydrogen, oxygen, and nitrogen important to life on Earth?

4. How can there be so many different substances in the world if there are only a few elements that are common?

Particles

Elements are the building blocks of matter. On Earth there are 90 different elements. Carbon is one element. Aluminum is another element. Gold is a third element. All of the elements are different from one another. Each element has its own unique properties.

What Are Elements Made Of?

If you cut a copper wire in half, you have two smaller copper wires. If you cut each of those short wires in half, you have four really short copper wires. If you cut those in half, you have some little bits of copper that don't look like wire anymore. But they are still pieces of the element copper.

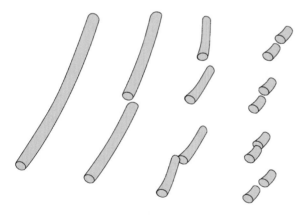

Copper wire cut into small pieces is still the element copper.

Imagine that you could cut one of those bits of copper into a million tiny pieces. Then cut one of those tiny pieces into a billion pieces. The pieces would be too small to see even with a microscope. But they are still pieces of the element copper.

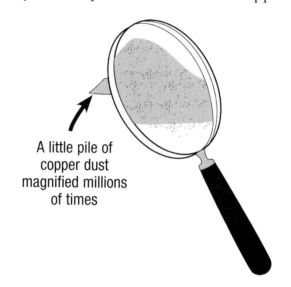

A little pile of copper dust magnified millions of times

If you keep cutting the copper bits in half, you will finally end up with the smallest possible piece of copper. The smallest piece of copper is the copper **particle**. If you cut the copper particle in half, it is no longer copper.

One copper particle magnified billions and billions of times

Particle is the word used to describe the smallest piece of any substance that is still that substance. Every substance has its own kind of particle. Copper is made of copper particles. Aluminum is made of aluminum particles. Gold is made of gold particles. Each of the 90 elements is made of its own kind of particle.

More Than 90 Particles

Remember, elements combine to make new substances. Elements can combine in millions of different ways to make millions of different substances. Each different substance has its own unique particle. So there are millions of different kinds of particles in the world.

Here is an example. Two hydrogen (H) particles and one oxygen (O) particle can combine. That forms the substance water. The chemical formula for water is H_2O. The formula shows that the water particle is made of two hydrogen particles and one oxygen particle.

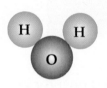
The elements hydrogen and oxygen combine to form a water particle.

Sodium bicarbonate is a substance. Like all substances, it has a particle. The chemical formula for sodium bicarbonate shows which elements combine to make a sodium bicarbonate particle. The chemical formula for sodium bicarbonate is $NaHCO_3$. Which element particles combine to make one sodium bicarbonate particle? One sodium particle, one hydrogen particle, one carbon particle, and three oxygen particles.

The elements sodium, carbon, oxygen, and hydrogen combine to form a sodium bicarbonate particle.

All of the other substances you have worked with have particles. The sodium chloride particle (NaCl) is small. Magnesium sulfate ($MgSO_4$) is a larger particle. And the sucrose particle ($C_{12}H_{22}O_{11}$) is the largest of the ones you used.

Representing Particles

Particles of substances can be lots of different shapes. Some particles are long and thin. Some are bumpy. But even the largest particles are far too small to see.

Even though particles are different sizes and shapes, you can represent them as little balls. You can draw the little balls in different colors when you want to show that particles of two or more substances are present. Representing particles as balls of different colors allows you to think about how particles are organized in a substance and what the particles are doing.

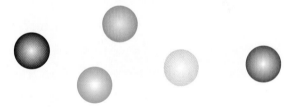

Review Questions

1. What is a particle?

2. What is the difference between an element and a particle?

3. How many different kinds of particles are there in the world? Explain your answer.

THREE PHASES OF MATTER

On an early spring day, Arturo went out to the coast. He walked out on a rocky point. A stiff, cold wind was blowing. Wave after wave crashed on the rocks. It was a great day to experience the three **phases** of matter.

What are the three phases of matter? **Solid**, **liquid**, and **gas**. The solid rocks provided a secure place to stand. The liquid water looked great as it flowed around the rocks and then retreated. And the invisible air was blowing by at a good rate. Solid, liquid, and gas matter made for a memorable day.

Rock, water, and air have different properties. Rock has definite size **(volume)** and shape. The size and shape don't change. That is characteristic of all solid substances.

Water has definite volume, but its shape changes. It forms waves, crashes up on shore, and flows back to sea. The volume of water is always the same, but its shape depends on where it is. That is characteristic of all liquid substances.

Air does not have definite volume or shape. Air fills in around everything and can take any shape. The ability to change volume and shape is characteristic of all **gaseous** substances.

Why is rock solid? Why is water liquid, and air gas? The phase of a substance depends on the relationship between the particles of that substance. In rock, the particles are stuck together. They can't move around. Attractive forces called **bonds** "glue" particles together. When particles can't move, the substance can't change size or shape.

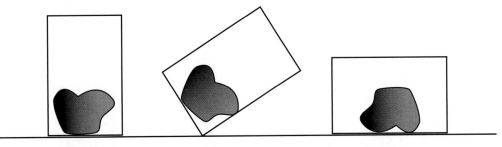

Solids have definite volume and shape and sit on the bottom of their containers.

In water, the particles are close to each other, but they are not stuck together. The particles move over and around each other. That allows water to flow. Particles in a liquid are packed close together, so the volume always stays the same. But, because particles can move, the shape of a liquid can change. Liquids take the shape of their containers.

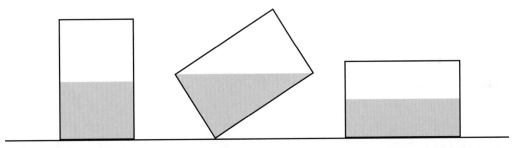

Liquids have definite volume, but their shape changes to fill the bottom of their containers.

In gas, particles fly around in space as individuals. There is a lot of space between particles. Because they are not attached, they spread out to fill a container. Gases have no definite volume or shape. Gases fill any container they are put into.

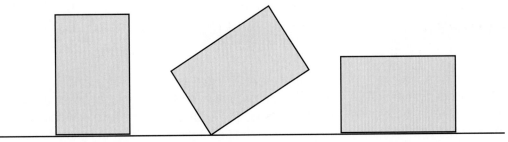

Gases have no definite volume or shape and always fill their containers.

Matter as Particles

Let's look at three samples of matter. One sample is solid, one is liquid, and the third is gas. Each sample is made of 39 particles. Each is placed in an identical container. This is how the particles would be organized in the three samples.

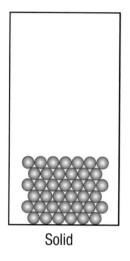
Solid

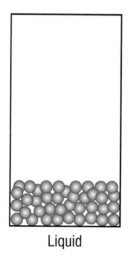
Liquid

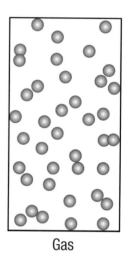
Gas

Applying Force to Matter

A syringe is a good tool for applying **force** to samples of matter. We can put a plunger into each of the cylinders above to see what happens when force is applied to a solid, a liquid, and a gas.

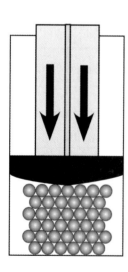

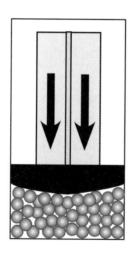

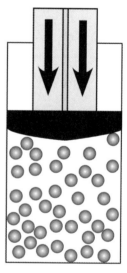

The particles of a solid are touching and bonded tightly. Force cannot change the shape of the solid or push the particles closer together.

The particles of a liquid are touching, but they can move. Force can change the shape of the liquid but cannot push the particles closer together.

The particles of a gas have a lot of space between them. Force can change the shape of the gas and can push the particles closer together.

Solids and liquids aren't very interesting in a closed syringe. They can't be forced into a smaller space, because the particles are already in contact with one another. But there is a lot of space between gas particles. Gas can be **compressed**. The particles of compressed gas are forced closer together.

But there is a limit to how much gas can be compressed. At first it is easy to push the plunger down. The air inside the syringe feels spongy. But the farther down you push the plunger, the harder the air feels. Why is that?

The air particles are flying around inside the syringe very fast. When they crash into the plunger and the walls of the syringe, they apply a force.

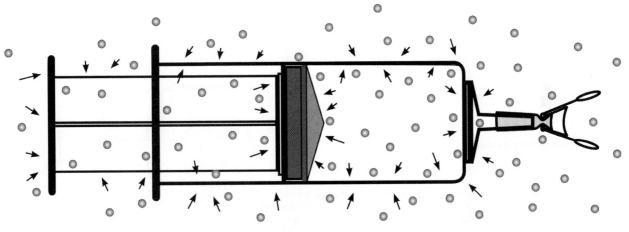

Moving air particles apply a force to everything they hit.

When gas is pushed into a smaller space, the concentration of particles increases. That results in more particles hitting the plunger every second. The particles pushing on the plunger tip are what makes it first hard and then impossible to push the plunger farther.

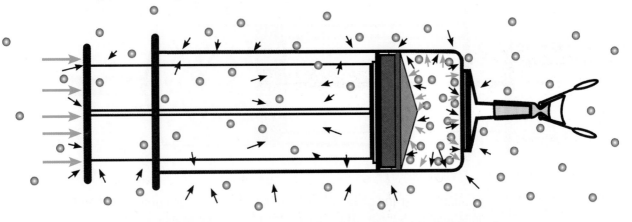

The force pushing the plunger down (blue arrows) is opposed by the force applied by the increased number of air particles (green arrows) hitting the plunger.

A Bubble in a Syringe

A plastic bubble filled with air crumples up when you compress the air around it. Can you figure out why?

When air is compressed, the air particles hit everything with more force. They hit the plunger, the walls of the syringe barrel, and the plastic surrounding the air in the bubble. The plastic pushes on the air inside the bubble. The air inside the bubble compresses, just like the air outside the bubble. The number of air particles inside the bubble stays the same. But the space occupied by those particles (the volume) is smaller. The plastic bubble crumples around the smaller volume of air inside.

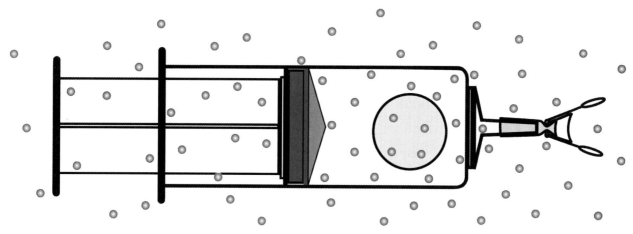

A syringe with an air bubble before the plunger is pushed in.

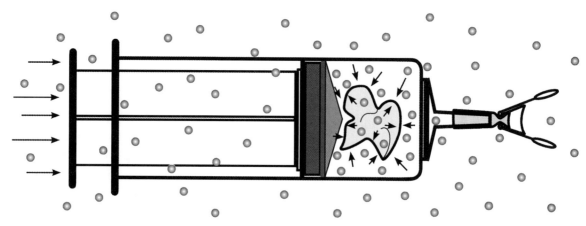

A syringe with an air bubble after the plunger has been pushed in. Air is compressed in the syringe and the bubble. This causes the bubble to crumple.

Foam Cubes in a Syringe

There are two kinds of foam rubber. The little spaces in the foam rubber are called cells. In open-cell foam rubber, all the cells are connected to each other and to the outside. In closed-cell foam rubber, the cells are not connected. Each cell is like a little bubble.

In open-cell foam, all the cells are connected.

In closed-cell foam, all the cells are separate.

Air particles in open-cell foam can move freely in and out of the cells. Air particles in closed-cell foam cannot move in or out of the cells.

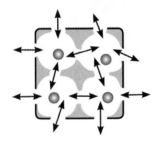

Air particles move easily through cells in open-cell foam.

Air particles cannot enter or leave cells in closed-cell foam.

Put an open-cell foam cube and a closed-cell foam cube in a syringe. The cubes are the same size. The air particles are the same distance apart in and around the cubes.

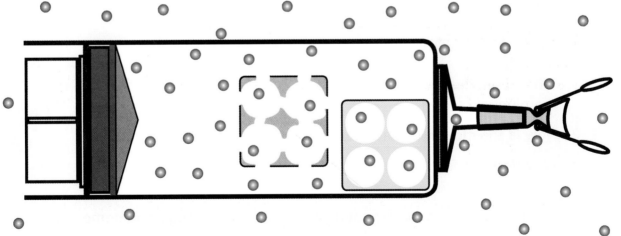

Air particles are evenly distributed in and around the cubes.
The cubes are the same size.

When you apply pressure to the plunger, the distance between the air particles in and around the cubes gets smaller. The distance is smaller because the air particles have been pushed closer together. And the two foam cubes no longer look alike.

The open-cell foam cube stays the same size. When the pressure increases, more air particles move into the open cells. When air is compressed, there are more particles inside the open-cell foam cube.

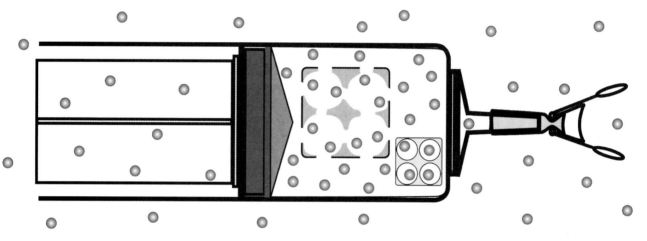

When force is applied to the syringe plunger, air particles are still evenly distributed in and around the cubes. But the particles are closer together. The cubes are not the same size.

The closed-cell foam cube gets smaller. Air particles can't enter the closed cells. The air particles outside the closed-cell cube push. They push on the cube until the distance between the particles in the cube is the same as the distance between the particles outside the cube. The only way this can happen is if the cells in the cube get smaller. Each tiny cell crumples and gets smaller, just like the plastic bubble.

Review Questions

1. What crumples a plastic bubble in a syringe when you apply force to the plunger?

2. How is the motion of particles in solid, liquid, and gas different?

3. Why does air feel hard when you push on the plunger of a closed syringe?

4. Explain why some foam cubes get smaller in a syringe and some stay the same size.

Particles in Motion

Air is matter. It has mass and occupies space. Air is a **mixture** of many gases. Air is approximately four-fifths nitrogen and one-fifth oxygen. All the other gases, including carbon dioxide and water vapor, make up only a little more than 1% of the mass of a sample of air.

Air is matter in its gas phase. That means that the nitrogen and oxygen particles in air are not connected to other particles. Gas particles fly through space as individuals.

same number of particles as every cubic centimeter of air outside the bottle.

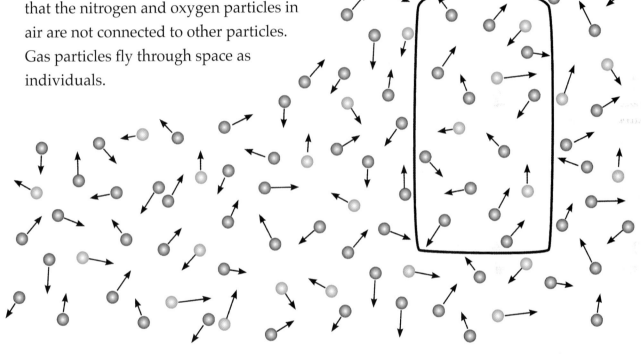

Air particles fly through space as individual particles. Air particles fill an open bottle.

After you drink a bottle of spring water, you have an excellent container for an air investigation. The empty bottle, of course, isn't empty. It is full of air. Because air particles are flying all around, they are going into and out of the open bottle all the time. The **density** of air in the bottle is exactly the same as the density of the air outside the bottle. That means that every cubic centimeter of air in the bottle has the

It is important to remember that air particles are really millions of times smaller than the representations in the illustrations. A cubic centimeter of air actually has about one quintillion air particles! A quintillion is a one followed by 18 zeroes (1,000,000,000,000,000,000). The illustrations are therefore not accurate, but they are good for thinking about what is going on at the particle level.

Particles Have Kinetic Energy

Not only are air particles incredibly small, they are always moving. And they move fast. At **room temperature**, they are going about 300 meters per second. That's equal to about 670 miles per hour.

Moving objects have energy. It's called **kinetic energy**. Anything that is in motion has kinetic energy, whether it is an ocean liner, a bicycle, a fly, a snail, you walking to class, water falling down a waterfall, or an oxygen particle in the air. They all have kinetic energy.

Kinetic energy, like all forms of energy, can do work. Air particles do work when they crash into things. Air particles push on each other, on you, on the walls of containers, and on everything else around them. Every air particle crashes into another particle about 10 billion times every second!

The amount of kinetic energy an object has depends on two things: the mass of the object and the speed at which it is moving. You can't change the mass of an air particle, but you can change its speed. By making a particle go faster, you increase its kinetic energy. Air particles can be made to move faster by heating a sample of air. Heat increases the kinetic energy of particles.

Back to the air investigation. Stretch a balloon over the top of the bottle full of air. Now the air is trapped inside the bottle-and-balloon system. No particles can get in or out.

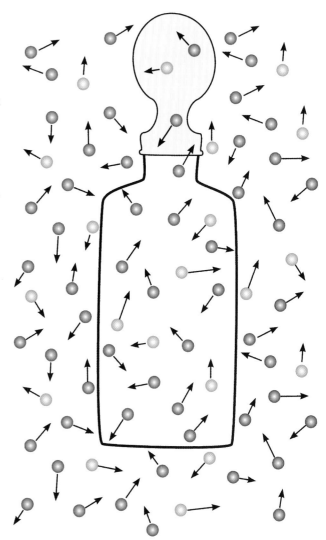

A balloon can trap the air inside a bottle.

The density of air particles is the same in the bottle, in the balloon, and in the air surrounding the bottle-and-balloon system.

Now place the bottle-and-balloon system in a cup of hot water. The hot water warms the air inside the bottle. Particles in the warm air start to move faster. After a few minutes, the bottle-and-balloon system looks like this.

Why did the balloon inflate? The hot water heated the air in the bottle. As a result, the air particles began moving faster. Faster-moving particles have more kinetic energy. Faster-moving particles hit each other harder, which pushes them farther apart. You can see in the illustration that the particles of warm air inside the bottle-and-balloon system are farther apart.

The faster-moving particles also push on the balloon membrane harder. The particles push hard enough to stretch the balloon membrane. The increased kinetic energy of the particles pushes them farther apart (air **expansion**), and the membrane stretches to hold the increased volume of air.

Hot water increases the kinetic energy of the air particles inside the bottle-and-balloon system. The particles fly faster and hit each other harder. The particles push farther apart, causing the gas to expand.

What Happens When Gases, Liquids, and Solids Heat Up?

Gas. If a sample of matter is gas, its particles are not bonded (attached) to other particles. Each particle moves freely through space. When a sample of air

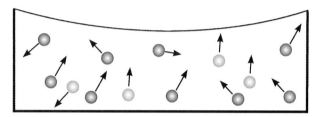

The particles in gases fly through space in all directions as individuals.

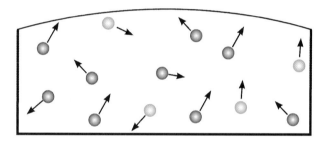

When gases get hot, the particles fly faster. Faster particles hit other particles harder, pushing the particles farther apart. This causes the gas to expand.

heats up, the particles move faster and hit each other harder. The result is that the particles push each other farther apart.

In the illustrations above, a container of gas has a flexible membrane across the top. When the gas gets warm, the kinetic energy of the particles increases, particles hit each other harder, and the gas expands. As the gas expands, it pushes the membrane upward.

Liquid. Particles in liquids are in close contact with one another. Attractions between the particles keep them from flying freely through space. The particles in liquids can, however, move over, around, and past one another. Individual particles in liquids are able to move all through the mass of liquid.

The particles in liquids are held close to each other. Particles bump and slide around and past each other.

When liquids get hot, the particles bump and push each other more. Increased bumping pushes the particles farther apart. This causes the liquid to expand.

The motion of particles in a liquid is kinetic energy. When a liquid gets warm, the particles move faster. The particles have more kinetic energy. As a result, they hit other particles more often and hit harder. This pushes the particles farther apart. When particles are pushed farther apart, the liquid expands.

Solid. Particles in solids have bonds holding them tightly together. The particles cannot move around at all. The particles are, however, still in motion. Particles in solids are always **vibrating** (moving back and forth) in place.

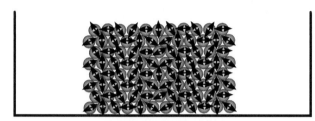

The particles in solids are bonded. Particles move by vibrating, but do not change positions.

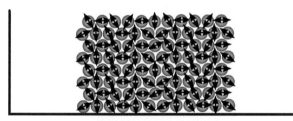

When solids get hot, the particles vibrate more. Increased vibration pushes the particles farther apart, causing the solid to expand.

The vibrational motion of particles in solids is kinetic energy. Heat makes the particles in a solid vibrate faster, giving them more kinetic energy. Faster-vibrating particles bump into one another more often and hit each other harder. This pushes the particles farther apart. When particles are pushed farther apart, the solid expands.

Summary

General Rule 1. When a sample of solid, liquid, or gas matter heats up, it expands. When matter gets hot, its particles gain kinetic energy. The increased kinetic energy pushes the particles farther apart. This causes the matter to expand.

General Rule 2. When a sample of solid, liquid, or gas matter cools down, it contracts. When matter cools down, its particles lose kinetic energy. The decreased kinetic energy lets the particles come closer together. This causes the matter to contract.

Review Questions

1. What is kinetic energy?

2. What are two ways to increase an object's kinetic energy?

3. Explain why a balloon inflates when a bottle-and-balloon system is placed in hot water.

4. What happens to a sample of matter when its particles lose kinetic energy?

5. How are particles in solids, liquids, and gases the same? How are they different?

Expansion and Contraction

We are surrounded by solids, liquids, and gases. And they all warm up and cool down from time to time. When they warm up, they expand. When they cool down, they contract. Sometimes expansion and **contraction** are useful. Other times they are a nuisance.

This bridge is made of solid steel with a solid surface of concrete. During the summer, the surface of the bridge heats up. The surface expands. When the surface gets bigger, what will happen to the bridge? If measures are not taken to allow the bridge to expand, it can buckle and break. Or the force of expansion may damage the steel structure.

Bridge engineers design the bridge's surface in sections. They place expansion joints where the sections meet. The bridge shown here has expansion joints with steel fingers that can move between each other. When the surface gets hot, the sections get longer and the fingers push between each other. When it gets cold, the sections get shorter, and the fingers pull apart. The finger design allows cars to cross the junctions between the surface sections smoothly.

This bridge has finger-type expansion joints between sections of its surface.

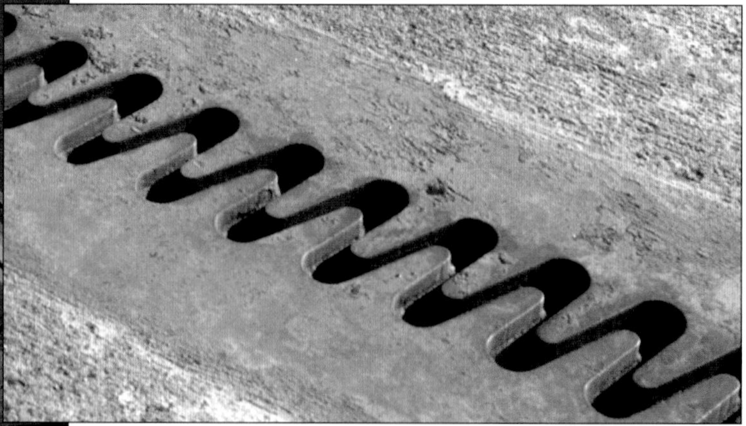

On a cold day, the bridge surface contracts. The fingers of the expansion joint don't overlap very much when the sections are short.

Thermometers are filled with liquid. The most commonly used liquid is alcohol. When the alcohol gets hot, it expands. When the alcohol gets cold, it contracts. Expansion and contraction of alcohol are useful properties for making a thermometer. Here's how it works.

If you want to know the temperature of the water in a vial, you place the bulb of a thermometer in the water. If the water is cold, heat moves from the warm thermometer bulb to the cold water. As the alcohol cools down, the kinetic energy of the alcohol particles declines. When the alcohol particles move less, they move closer together, and the alcohol contracts. When the alcohol contracts, it take up less space inside the thermometer. The level in the stem goes down.

When the thermometer is moved to a vial of warm water, the alcohol in the bulb heats up. The kinetic energy of the alcohol particles increases, and the volume of alcohol expands. The larger volume pushes farther up the thermometer stem.

By using the numbers on the thermometer stem, you can compare temperatures. These thermometers are **calibrated** in degrees Celsius (°C). The cool water in the picture is about 16°C. The warm water is about 51°C.

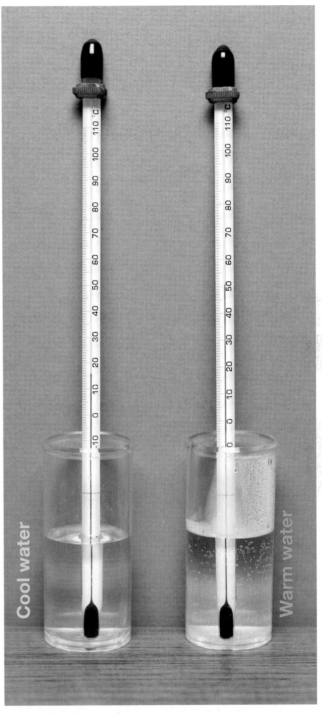

A thermometer contains alcohol, which expands when it gets warm and contracts when it gets cool. The height of the alcohol column indicates the temperature of the water around the thermometer bulb.

Did you ever see a container of fruit salad open itself? Here's the chain of events that can make that happen.

You mix up some grapes, strawberries, and melon. You eat some of the salad, but there is some left for a snack later. Snap a tight-fitting plastic lid on the container of fruit salad and pop it in the refrigerator to keep it fresh.

A container of fruit just out of the refrigerator

Later you take the salad out of the refrigerator and set it on the counter. If you don't open the container right away, in 10 or 15 minutes the lid pops off the container, reminding you to enjoy your snack. Why?

The air inside the container is gas. As the container of fruit salad warms up, so does the air in the container with the fruit. The kinetic energy of the air particles increases, and they push harder on each other and on the lid of the container. If you look closely, you can see the lid bulging as the air warms. Eventually the pressure created by the increased kinetic energy of the air particles will pop the lid right off the container.

The container lid bulges upward after 10 minutes.

Expansion and contraction occur around us all the time. Air expands and contracts as it warms and cools. We don't see it happening, but we see the results. Weather, particularly wind, happens as a result of expanding and contracting air.

When the catch is released, the lid pops off.

Have you ever been in an old, quiet building when there was a sudden creaking or cracking sound? Is it a haunted house? Probably not. Buildings expand and contract as they heat and cool. When they do, beams, walls, and roofs move a little bit. When parts of a building move, they can make sounds. Listen closely next time you are in a quiet building just after sunrise or just after sunset. Those are the times when you are most likely to hear unexpected sounds. Do you know why?

Train tracks are steel. Tracks expand when they get hot in the sunshine. What keeps them from buckling and bending? Expansion joints. Next time you are close to some train tracks, look for the expansion joints. What keeps sidewalks and roads from breaking and buckling? Expansion joints. These joints are spaces between the concrete sections. The spaces are filled with material that can compress when the concrete expands.

Keep your eyes open when you are looking at very large structures and huge expanses of solid surface. If you look closely, you will probably see how the engineers solved the problems presented by expansion and contraction caused by heating and cooling.

Review Questions

1. What are expansion joints, and why are they used?

2. What causes the cap to pop off a bottle of orange juice?

3. How does a thermometer work?

Train track expansion joint

Compressible concrete joint

Energy on the Move

Everybody knows that a cup of too-hot cocoa can be cooled to sipping temperature by adding a splash of cold milk.

When you mix cold milk and hot cocoa, what happens to the cold milk? And what happens to the hot cocoa?

Kinetic Energy = Heat

Objects in motion have kinetic energy. Particles that make up substances are objects. Particles are always moving. Therefore, the particles that make up the hot cocoa and the cold milk have kinetic energy.

Cold milk being poured into hot cocoa

The amount of energy a particle has depends on how fast it is moving. Faster-moving particles have more energy. Slower-moving particles have less energy.

The particles in hot substances have more kinetic energy than the particles in cold substances. The simple rule is that the more kinetic energy the particles have, the hotter the substance is.

After mixing, the cup of cocoa is warm throughout. It is cooler than the hot cocoa and hotter than the cold milk. It seems as if the hot cocoa gets colder and the cold milk gets hotter. The new temperature is between the starting temperatures of the cocoa and the milk.

How does that happen?

This is important. Kinetic energy of particles is directly related to heat.

Changing Kinetic Energy

Energy is **conserved**. That simply means that energy is never destroyed or created during interactions. The amount of energy in a system is always the same, but it can move from place to place. When energy moves from one place to another, it is called an **energy transfer**.

Energy transfer happens when particles collide. When a fast-moving particle hits a slow-moving particle, the slow-moving particle speeds up. The fast-moving particle slows down. When a particle speeds up, it has more kinetic energy. When a particle slows down, it has less kinetic energy. Energy transfers from a fast-moving particle to a slow-moving particle at the moment of impact. Visualize this situation.

1. Fast-moving particle 1 is on a collision course with slow-moving particle 2.

2. The particles collide. At the moment of impact, energy transfers from particle 1 to particle 2. As a result, particle 1 has less kinetic energy, and particle 2 has more kinetic energy.

3. The two particles are now moving at about the same speed. Energy transferred from particle 1 to particle 2 at the moment of impact.

If you add up the kinetic energy of the two particles before the collision, it is exactly the same as the kinetic energy of the two particles after the collision. No kinetic energy is created or lost as a result of the collision. The energy of the two-particle system is conserved.

But something did change. As a result of the collision, particle 1 has less kinetic energy, and particle 2 has more kinetic energy. The collision resulted in an energy transfer. Energy transferred from particle 1 to particle 2.

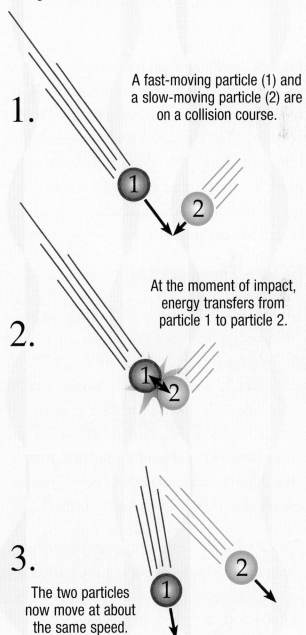

1. A fast-moving particle (1) and a slow-moving particle (2) are on a collision course.

2. At the moment of impact, energy transfers from particle 1 to particle 2.

3. The two particles now move at about the same speed.

Energy Transfer in Cocoa

The kinetic energy of the particles in hot cocoa is high. They are all moving fast. The kinetic energy of the particles in cold milk is low. They are all moving slowly.

When you pour cold milk into hot cocoa, the milk and cocoa particles start to collide. When a high-energy cocoa particle hits a low-energy milk particle, energy transfers. The cocoa particles slow down, and the cup of cocoa cools down.

Look at the illustration on page 33. Can you tell which of the particles represents a cocoa particle and which represents a milk particle? And can you see how the energy transfer reduced the speed of the cocoa particle? Remember, reduced particle speed means less kinetic energy. Lower kinetic energy means lower temperature.

Using a Thermometer

You can use a thermometer to find out if your cocoa is too hot. When you dip a thermometer in that cup of hot cocoa, it reads 90°C. Whoa, too hot. Need to add some cold milk. Move the thermometer over to the cold milk, and the thermometer reads 10°C. That should do the job. But how does the thermometer "know" that the cocoa is 90°C and the milk is 10°C?

Kinetic energy. The thermometer reports the **average kinetic energy** of the particles in a substance. That's what temperature is: average kinetic energy of particles.

How exactly does the thermometer work? Now that we have an idea of how energy transfer takes place, let's put the thermometer into the 90°C cocoa. The cocoa particles collide with the glass particles on the outside of the thermometer stem.

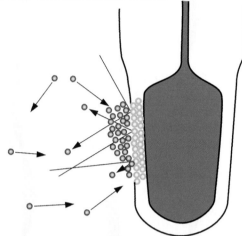

Energy transfers from cocoa particles to glass particles.

The glass particles gain kinetic energy and start vibrating more rapidly. The glass particles transfer energy to their neighbors, and those transfer energy to their neighbors. Transfer of energy from particle to particle by contact is called **conduction**. Pretty soon the whole glass stem is at 90°C.

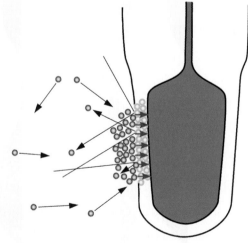

Energy transfers through the glass by conduction.

The rapidly vibrating glass particles that are in contact with the alcohol inside the thermometer stem transfer kinetic energy to the alcohol particles. Kinetic energy is conducted from alcohol particle to alcohol particle.

In 90°C cocoa, the alcohol rises to the 90°C mark on the thermometer.

In 10°C milk, the alcohol rises to the 10°C mark on the thermometer.

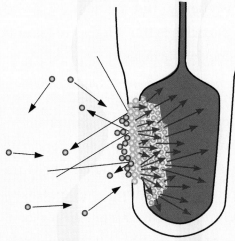

Energy transfers from the glass particles to the alcohol particles. The alcohol expands.

Soon all the alcohol particles are moving faster, pushing on each other more often and with greater force. The distance between particles increases, and the alcohol expands. As the alcohol expands, its volume increases. Alcohol pushes up the stem of the thermometer. The greater the kinetic energy of the alcohol particles, the more the alcohol expands. Energy transfers to the alcohol until the average kinetic energy of the alcohol particles is the same as the average kinetic energy of the cocoa particles. When this happens, the top of the alcohol is at the 90°C mark.

Put the thermometer into the 10°C milk. Now energy transfers from the outside of the glass thermometer stem to the milk. Energy transfers from glass particles, lowering the kinetic energy of all the glass particles. When alcohol particles collide with lower-energy glass particles, energy flows from the alcohol to the glass. The alcohol loses kinetic energy and contracts. When the average kinetic energy of the alcohol particles is the same as the average kinetic energy of the milk particles, the alcohol level is at the 10°C mark.

Energy Flows from High to Low

When two particles collide, is it possible for the fast-moving particle to end up going even faster? Can energy transfer from a lower-energy particle to a higher-energy particle?

No. It never happens. Energy *always* transfers from a faster-moving particle to a slower-moving particle of the same size. As a result of an energy-transfer collision, the particle that was going faster before the collision will always be going slower after the collision. And the particle that was going slower before the collision will be going faster after the collision. Always.

It is sometimes useful to think of energy as flowing. Energy always flows from higher to lower, from hot to less hot (cold).

When Does Energy Stop Flowing?

When you pour cold milk into the hot cocoa, the high-energy cocoa particles and low-energy milk particles instantly mix with one another. They collide with each other billions of times every second. The energy flows from the high-energy cocoa particles to the low-energy milk particles.

Almost instantly the average kinetic energy of the milk particles is the same as the average kinetic energy of the cocoa particles. The kinetic-energy level is uniform throughout—lower than the cocoa and higher than the milk.

Has energy stopped flowing? Has energy transfer between particles stopped? No, not really. Even when the average kinetic energy of the mixture stays steady, there are still individual particles that have high energy, and there are particles that have low energy. But the number of high-energy particles is the same as the number of low-energy particles.

When the temperature is constant, the system is in a condition called **equilibrium**. At equilibrium, there is no change of temperature. When a mixture of hot cocoa and cold milk has reached equilibrium, you can use a thermometer to measure the equilibrium temperature. The equilibrium temperature is a measure of the average kinetic energy of all the particles in the system. This includes the cup, the mixture of cocoa and milk, and the thermometer.

So, has energy stopped flowing? Think about this. The phone rings and you talk to a friend for 20 minutes. Now the cup of cocoa is cold. Why? The room is cooler than the cocoa. Particles of air collide with the cup and the surface of the cocoa. Energy transfers from the cup of cocoa to the air. Energy continues to transfer to the air until the average kinetic energy of the cocoa is the same as the average kinetic energy of the air. We say the cocoa is room temperature. The cocoa is at equilibrium with everything else in the room.

Energy Transfer Summary

All matter is made of tiny particles that are too small to see. The particles are in constant motion.

Objects in motion have kinetic energy. Particles are objects in motion, so they have kinetic energy. The faster a particle moves, the more kinetic energy it has.

Kinetic energy is related to heat. The faster the particles in a substance move, the hotter it is.

Energy can move, or transfer, from one particle to another when particles collide. Energy always transfers from a higher-energy particle to a lower-energy particle. The transfer of kinetic energy from particle to particle as a result of contact is called conduction.

Temperature is a measure of the average kinetic energy of the particles in a mass.

Matter heats up and cools down because of energy transfer at the particle level.

Review Questions

1. Explain how cold milk cools hot cocoa.

2. Why do you think an ice cube feels cold when you hold it in your hand?

3. What will happen to the balloon stretched over the mouth of this "empty" bottle when the bottle is placed in hot water? Explain all the energy transfers.

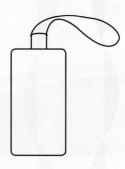

4. When does energy flow from a cold object to a hot object?

5. What does a thermometer measure, and how does it do it?

HEAT OF FUSION

You have probably had this experience a hundred times. It's a warm day, and you are at a party or a picnic. You are thirsty, so you go to the ice chest and take out an ice-cold soft drink. You pop the top and take a swallow. So cold and refreshing! One of life's small pleasures.

The ice chest is a great invention. It is simple to maintain, easy to use, and very efficient for cooling drinks. But did you ever think about how it works? It is not quite as simple as it seems.

HOW DOES ICE MAKE THINGS COLD?

If you ask your little brother how an ice chest works, he might have this idea.

Ice is cold. When you put it in an ice chest, it makes the whole inside of the chest cold. When you put drinks in with the ice, the cold goes into the soft drinks. Cold is stronger than heat, so it just takes over. That's why drinks get cold in an ice chest.

That's an interesting story, but it is not going to work. We now know that there is

38

no such thing as cold. There is only energy of particles. The greater the energy, the hotter the material. What we sense as cold is actually a low level of particle energy. Objects get cold when energy transfers from them to another location. Soft drinks in an ice chest get cold because energy transfers from the soft drinks to something else. The "something else" in the ice chest is the ice itself. And what happens to the ice? It melts.

What Really Happens When Ice Melts?

Boiling water is 100°C. Ice that has just formed is 0°C. Ice that has just melted is 0°C, too. If we mix 100 grams (g) of 100°C water with 100 g of 0°C water, in a moment the mixture will reach equilibrium. We will have 200 g of water at 50°C.

The equation that lets us predict the final temperature when we mix equal volumes of water of two different temperatures is this

$$T_f = \frac{T_h + T_c}{2}$$

T_f = final temperature
T_h = temperature of hot water
T_c = temperature of cold water

We can use the equation to predict that the final temperature of the mixture described above will be 50°C.

$$T_f = \frac{T_h + T_c}{2} = \frac{100°C + 0°C}{2} = 50°C$$

Now let's mix 100 g of 100°C water with 100 g of 0°C ice. The starting temperatures are 100°C and 0°C. So we might predict that the final temperature of the mixture when it comes to equilibrium will again be 50°C. But it is not. The equilibrium temperature is only 10°C. Why?

Energy transfers from particles in the hot water to particles in the cold water.

Energy transfers from particles in the hot water to particles in the ice.

It takes a lot of energy to melt ice. Ice is the solid phase of water. Water particles in ice are held in place by forces called bonds. In order for solid water to turn into liquid water, the bonds must be broken. It takes energy to break bonds. That's where most of the energy transferred from the hot water goes. The energy changes water from a solid to a liquid.

But the energy that melts the ice does not change its temperature. Ice is 0°C, and the liquid water it turns into is also 0°C. The energy that transfers to ice doesn't change the kinetic energy of the water particles, it just breaks bonds. The energy that breaks bonds to change solid water into liquid water is called **heat of fusion**.

CALCULATING HEAT OF FUSION

We can calculate the heat of fusion using the equation for calculating **calories**. First we calculate the heat transferred from the hot water (cal_h).

$$cal_h = \Delta T \times m$$

$\Delta T =$ temperature change
$m =$ mass

$$cal_h = 90°C \times 100 \text{ g} = 9,000 \text{ cal}$$

Now we calculate the heat transferred to the ice (cal_c).

$$cal_c = \Delta T \times m$$
$$cal_c = 10°C \times 100 \text{ g} = 1,000 \text{ cal}$$

It appears as though the number of calories transferred from the hot water exceeds the number of calories transferred to the cold ice by 8,000 calories. But we know that energy is conserved. No energy is ever created, destroyed, or lost during energy transfers. So what happened to the 8,000 calories?

The 8,000 calories were used to break the bonds to melt the 100 g of ice. We can do one last calculation to find out how many calories are needed to melt just 1 g of ice. Divide the total energy by the number of grams of ice in the sample.

$$\frac{8,000 \text{ cal}}{100 \text{ g}} = 80 \text{ cal/g}$$

The heat of fusion for water is 80 cal/g. That means for every gram of ice that melts, 80 calories of heat transferred from someplace to make that happen.

USING HEAT TO MAKE THINGS COLD

Heat of fusion is what makes the ice chest so good at cooling those soft drinks. The best way to set up a cooler is to fill it about half full with crushed ice. Then pour in just enough water to float the ice. The water will transfer energy to the ice, and some of it will melt. But as soon as the water gets down to 0°C, ice will stop melting. Why? Because the kinetic energy of the particles in the 0°C water will be the same as the kinetic energy of the particles

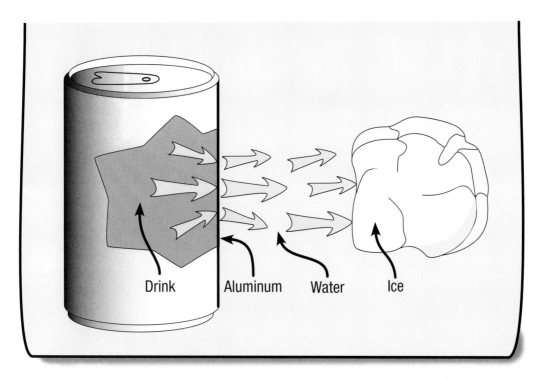

Energy transfers from drink particles to aluminum particles. The drink cools down.
Energy transfers from the aluminum particles to water particles. The aluminum can cools down.
Energy transfers from water particles to ice particles. Water cools down and ice melts.

in the 0°C ice. No more energy transfer. The system is in equilibrium at 0°C.

Now in go the room-temperature soft drinks. They sink down into the ice and water. Energy starts to transfer from the surface of the 20°C cans to the 0°C water surrounding them. As a can loses kinetic energy, the drink inside the can starts to transfer energy to the can. The water warms up, and the can and the drink inside cool down.

But the water doesn't stay warm. It transfers energy to the ice. Ice melts, keeping the temperature of the water at 0°C. Energy continues to transfer from the soft drinks and then to the ice until everything in the cooler is at 0°C.

What has changed? Energy has moved from the soft drinks to the ice. The drinks are cold, and some of the ice has melted. The ice chest system uses kinetic energy from the soft drinks to melt ice. The result is cold drinks. The ice chest uses energy transfer to cool the drinks.

Review Questions

1. What is heat of fusion?

2. What happens at the particle level when you put ice cubes in a glass of room-temperature lemonade?

3. Explain how an ice chest cools a can of soft drink.

ROCK SOLID

WHAT DOES LAVA POURING OUT OF A VOLCANO HAVE IN COMMON WITH A SNOWMAN?

They are both going to change phase in a short time. The liquid lava will **freeze** and become solid rock. The solid snowman will **melt** and become liquid water.

Most matter on Earth exists in one of three forms, solid, liquid, or gas. The forms are called states or phases of matter.

The clothes you wear, the forks and spoons you eat with, and your books and pencils are a few examples of matter in its solid phase.

The olive oil you put on your salad, the shampoo you use to wash your hair, and a refreshing glass of cold milk are examples of matter in its liquid phase.

The helium in a party balloon, the air you pump into a soccer ball, and the carbon dioxide in your exhaled breath are examples of matter in its gas phase.

PROPERTIES OF THE PHASES OF MATTER

Many substances can exist in more than one phase. The snowman, for instance, is made of solid water. We have many names for solid water, including ice, frost, and snow.

Water can also exist as liquid. Liquid water falls from clouds as rain and flows to your home in pipes. Earth has an ocean filled with liquid water.

Water also exists as gas. Water in gas phase is called **water vapor**. We are usually not aware of water vapor because it is invisible. Most of the water vapor on Earth is in the atmosphere as part of the air.

Ice, liquid water, and water vapor all look different. But they are all forms of water. What is the same and what is different about ice, water, and water vapor?

All three phases of water are made of exactly the same kind of particle. The chemical formula for the water particle is H_2O. Ice, water, and water vapor are all made of H_2O particles.

The thing that is different about ice, water, and water vapor is the relationship between the water particles.

In the article called *Three Phases of Matter*, we described how solids, liquids, and gases differ. In solids, the particles are attached to one another. The attachments are called bonds. The bonds in solids are so strong that the particles cannot change positions. That's why solids have definite shape and volume.

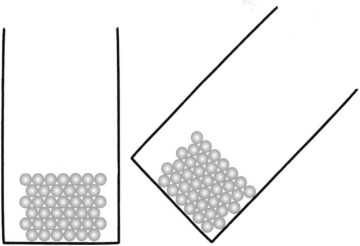

Ice in a vial will move from side to side, but will not change volume or shape.

43

In liquids, the bonds are weaker. The particles are still held close together, but they can move around and past one another. As a result, liquids flow. That's why liquids have definite volume, but their shape changes.

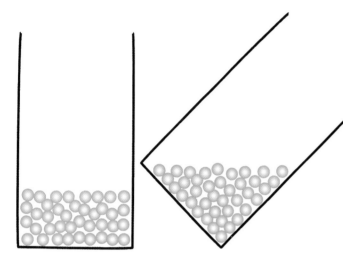

Liquid water has definite volume, but its shape changes to fit the container it is in.

In gases, the particles are not held together by bonds. Individual particles of gas fly around in space. That's why gases do not have definite volume or shape.

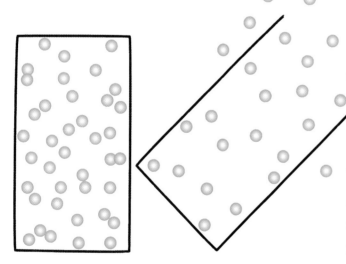

Water vapor does not have definite volume or shape. If the container is open, the gas will expand, and the particles will leave the container.

PHASE CHANGE

The snowman wasn't always solid. And it won't stay solid. The solid snowman will melt and turn to liquid water. The liquid **lava** wasn't always liquid. And it won't stay liquid. The liquid lava will freeze and turn to solid rock.

Change from solid to liquid, and change from liquid to solid, are examples of phase change. What causes substances to change phase?

Heat causes phase change. Or, more accurately, energy transfer causes phase change. Here's how it works.

When a piece of ice is placed in a warm room, energy transfers from the air particles to the water particles in the ice. The kinetic energy of the water particles increases until the ice reaches 0°C.

As more and more energy transfers to the 0°C ice, the bonds holding the water particles together start to break. When most of the bonds are broken, the water particles are no longer held in place. They start to move over and around one another.

When particles flow over and around one another, we say the substance changed from solid to liquid. The process is melting. Substances melt when enough energy transfers to the particles of a solid.

44

That's why the snowman melts. Energy from the Sun transfers to the water particles in the snow crystals. The bonds holding the snow crystals together break, and the solid water changes to liquid water. The snowman changes into a hat and scarf on top of a puddle of water.

What about the lava? How does it change phase? When lava pours out on Earth's surface, it is extremely hot. The kinetic energy of the rock particles is so great that most of the bonds holding them together have been broken. The rock particles move over and around one another. The rock is liquid, and it flows down the side of the volcano.

Air is cooler than lava. Energy from the rock particles transfers to the air particles. The rock particles lose kinetic energy, and the mass of lava cools. As the lava cools, bonds form between the rock particles. When enough energy has transferred from the rock particles, strong bonds form and the particles are locked in place.

When particles stop flowing over and around one another, we say the substance changed from liquid to solid. The process is freezing. Substances freeze when enough energy transfers away from the particles of a liquid.

That's why the liquid lava freezes and becomes rock solid. Energy transfers from the rock particles, bonds form, and the rock changes from liquid to solid.

MORE HEAT

What happens to the snowman next? After a day or two, all that remains is the hat and scarf. Even the puddle of liquid water has disappeared. Where did the water go?

As sunshine falls on the puddle of liquid water, energy transfers to the water particles. The kinetic energy of the particles increases. When enough energy transfers to a particle, the particle breaks all the bonds holding it to the mass of liquid. The particle breaks free and flies into space. The water changes phase again, but this time from liquid to gas.

The phase change from liquid to gas is called **evaporation** (or vaporization). Water in the gas phase is called water vapor. The individual water particles are too small to see, so water vapor is invisible. Water vapor enters the air and becomes part of Earth's atmosphere.

Water can change from gas to liquid, too. The process involves energy transfer. Can you predict where the energy transfer takes place?

When energy transfers from the water vapor particles, they lose kinetic energy. When enough energy has transferred away from the particles, bonds form between them. The water changes phase from gas to liquid. The process is **condensation**. Substances condense when energy transfers away from the particles of a gas.

Below is an illustration of an experiment similar to the one you did in class. A pan of liquid water is heated. Water evaporates. The water vapor condenses on a cup filled with ice. Study how the water particles change phase from liquid to gas, and then back to liquid. You should be able to see where evaporation and condensation are taking place.

MELT AND FREEZE

There are three important things to understand about melting and freezing.

Substances don't have to be cold to freeze. *Freeze* just means changing phase from liquid to solid. Granite freezes at about 1,650°C. On the other hand, oxygen freezes at –218°C. Every substance has its own freezing temperature.

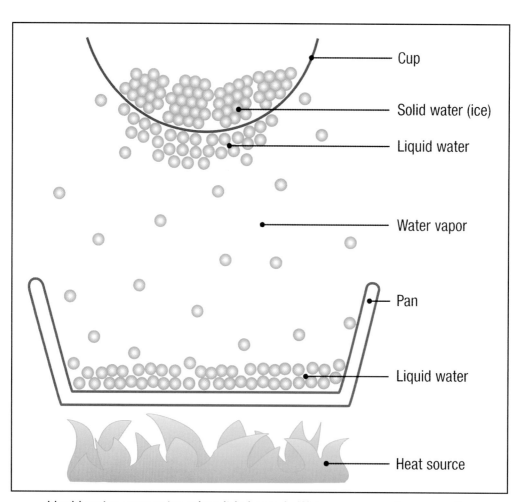

Liquid water evaporates when it is heated. Water vapor condenses on the outside of a cup containing ice.

Phase is a relationship between particles. The phase of a substance is determined by what is happening between the particles in the substance. Particles in solids have strong bonds, particles in liquids have weak bonds, and particles in gases have no bonds.

Freezing temperature = melting temperature. A substance freezes and melts at the same temperature. Water, for instance, freezes and melts at 0°C. If you move a piece of ice from a freezer to a warm room, the ice will warm up until it reaches 0°C. Then it will melt. If you put a cup of warm water in a freezer, the water will cool until it gets to 0°C. Then it will freeze.

The temperature at which a substance evaporates is the same as the temperature at which it condenses. Water, for instance, evaporates and condenses at 100°C.

Substance	Freeze/melt (°C)	Condense/evaporate (°C)
Helium	−272	−269
Oxygen	−218	−183
Nitrogen	−210	−198
Carbon dioxide	—	−78
Chlorine	−101	−34
Mercury	−39	357
Water	0	100
Sodium	98	883
Lead	327	1,749
Aluminum	660	2,519
Calcium chloride	775	1,936
Sodium chloride	801	1,465
Silver	962	2,162
Gold	1,064	2,856
Copper	1,085	2,562
Iron	1,538	2,861
Tungsten	3,422	5,555

The freeze/melt temperatures and condense/evaporate temperatures for some common substances

This illustration summarizes how energy transfer affects phase change. The top half shows how substances go from solid to liquid to gas as energy transfers to the particles of the substance. The bottom half shows how substances go from gas to liquid to solid as energy transfers from the particles of the substance.

Notice that it is possible for a substance to go straight from solid to gas. Carbon dioxide is an example of a substance that **sublimes**. And, when energy transfers from carbon dioxide gas, it **deposits** as solid without going through a liquid phase. Solid carbon dioxide is called **dry ice**.

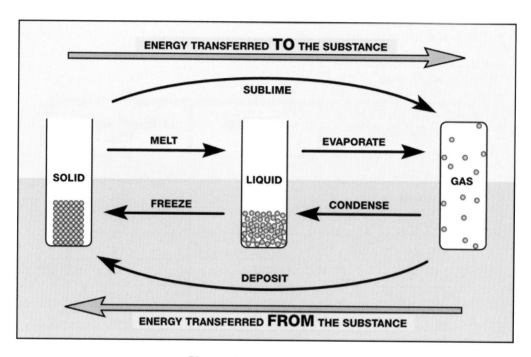

Phase-change vocabulary

REVIEW QUESTIONS

1. What causes a substance to change from one phase to another?

2. What are the three important things to know about freezing and melting?

3. Why does liquid water form on the bottom of a cup of ice placed over warm water?

4. What happens to water particles as a cup of ice melts and then evaporates?

How Things Dissolve

In the winter, you might get a cold. The sore throat and cough are very unpleasant. To get relief, you can pop a cough drop in your mouth. The sweet candylike cough drop contains medicine that makes your throat feel a little better and puts the cough on hold for a while.

But what happens to the cough drop? After a few minutes, it is a lot smaller than it was when you put it in your mouth.

or two, you can see a pool of red color forming around the cough drop. The red color is coming from the solid cough drop. Is the cough drop melting? Is it changing from solid to liquid?

No, the cough drop is not melting. Melting is caused by heat. Heat is not transferring to the cough drop in the cup of water.

And not long after that, it is gone. Where did it go?

The cough drop, which is mostly sugar, **dissolved**. It dissolved in saliva, which is mostly water. As it dissolved, the medicine flowed down your throat, bit by bit, soothing the pain.

Cough Drops in Water

You can observe the dissolving process more easily by putting a red cough drop in a cup of plain water. Within a minute

The cough drop is dissolving. It is breaking apart bit by bit, and the bits are moving into the water. In 10 minutes, the cough drop will be very small, and there will be a thin layer of red water on the bottom of the cup. In 20 minutes, the cough drop will be gone, and the red layer at the bottom of the cup will be larger. A day later the red color will have moved higher in the cup, and in 4 days, the whole cup will be pink. The cough drop will be spread evenly throughout the water in the cup.

A Close Look at Dissolving

Water is made of water particles. They are in constant motion, bumping and banging around and over one another. And they bump and bang into everything that is in the water. If you put a cough drop in a cup of water, water particles will bang into it billions of times every second.

The sugar particles leave the cough drop and form bonds with a few water particles. The tiny sugar-and-water groups move off into the mass of water.

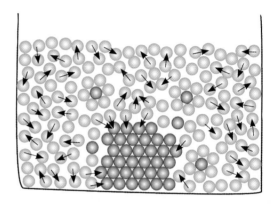

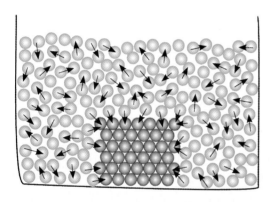

Cough-drop particles (shown in red) are held together by attractive forces called bonds. The bonds keep the cough drop from falling apart in the package before you use it. Water particles (shown in blue) hit the sugar particles on the edge of the cough drop and break the bonds.

After 20 minutes, the whole cough drop has been broken into particles and carried into the water. The cough drop is completely dissolved. And it happened one particle at a time.

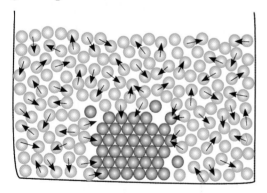

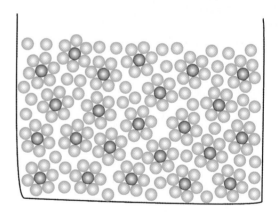

Mixture and Solution

When two or more materials are put together, you have a mixture. Peanuts and raisins make a good mixture for snacking. Tomato sauce and herbs is a mixture poured over pasta. A mixture of oil and vinegar is a good start for salad dressing. Any combination of materials is a mixture.

Salt dissolves in water. Salt dissolves just like the cough drop. After the salt has dissolved, the salt is no longer visible. The mixture is **transparent**. When two substances are mixed and one dissolves in the other, the mixture is a **solution**.

Solutions have two parts. There is the part that dissolves (salt) and the part that the salt dissolves in (water). The part of the solution that dissolves is the **solute**. The part that does the dissolving is the **solvent**.

Solutions on Earth

Remember this picture from the article called *Elements in the Universe*? In that article, we looked at the parts, called elements, that make up the ocean and the atmosphere. We discovered that the ocean is mostly hydrogen and oxygen, and the atmosphere is mostly nitrogen and oxygen.

Let's look at the ocean and atmosphere again, but this time to see how those parts are organized.

The ocean and the atmosphere

What is the largest solution on Earth? That's a tough question. Largest in volume or largest in mass? The most massive solution has to be the ocean. The ocean's depth ranges from a few meters (m) to over 10 kilometers (km) (6 miles). And the ocean covers more than 70% of Earth's surface. That's a lot of seawater.

Seawater is a solution. The solvent in seawater is plain water, the same kind of water you drink from the school drinking fountain. But seawater contains a lot of solutes. The main solute is the salt sodium chloride, the same salt you sprinkle on food. There are thousands of other solutes in seawater, too. Every element that occurs naturally on Earth is found in some quantity in seawater.

The solution that is largest in volume is Earth's atmosphere. The atmosphere covers Earth's entire surface, land and sea, and extends up about 600 km (375 mi.).

Earth's atmosphere is a mixture of gases called air. Air is pretty uniform in composition. It's about 78% nitrogen, 21% oxygen, 1% argon, and traces of hundreds of other substances. Does it seem a little odd to think of air as a solution? What is dissolved, and what did the dissolving? The substance that is present in greatest quantity is considered the solvent. So in air, nitrogen is the solvent. Oxygen, argon, carbon dioxide, water vapor, and all the other gases are solutes.

Solutions for Life

You are full of solutions. Water is the solvent for most of them. Saliva is a solution. So are sweat, urine, stomach acid, and tears. Each solution has an important function in the successful operation of a living human being.

Perhaps the most important solution in your body is blood. Actually, blood is a complex mixture. If you spin a sample of blood in a centrifuge, the solid parts of the blood will be forced to the bottom of the test tube. On top will be a clear, amber solution called **blood plasma**. Plasma is the solution. The solid portion of blood is mostly red and white blood cells. So blood is really a mixture. It's a solution with solids suspended in it.

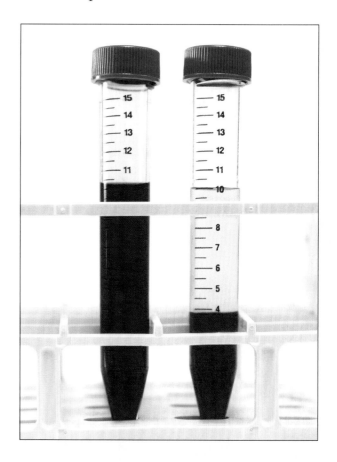

Plants contain a variety of solutions in their stems, leaves, and flowers. The sweet nectar that attracts bees and butterflies is a sugar solution. The sap flowing in the **phloem** of stems and leaves is a solution of sugars and salts. The solution drawn up the **xylem** tubes from the roots to all the plant's cells is a solution of minerals from the soil.

In every case, water is the solvent for life, and the solutes provide the raw materials that make life possible.

Solution Defined

- A solution is a mixture of two (or more) substances.

- In a solution, one substance (solute) is dissolved in the second substance (solvent). Solutions made with water as the solvent are transparent.

- In a solution, the solvent particles hit the particles of the solute and break the bonds holding the solute particles together. This is called dissolving.

- Substances that dissolve are **soluble**. Salt is soluble in water. Substances that don't dissolve are **insoluble**. Sand is insoluble in water.

- Individual solute particles are carried into the mass of solvent by the solvent particles. When the solute is all dissolved, the solute particles are uniformly distributed among the particles of the solvent.

- When one substance dissolves in another substance to make a solution, the particles of the two substances do not change. Solutions can be separated into their original substances. The most common way to separate a solution is by evaporating the solvent, which leaves the solute behind.

Review Questions

1. Copper chloride ($CuCl_2$) dissolves in water. Describe what happens at the particle level when copper chloride is put into water.

2. What are some of the solutions found in living organisms?

3. Is milk a mixture, a solution, or both? Why do you think so?

4. How could a solution of copper chloride and water be separated into its starting substances?

Concentration

If you stop by the freezer section at the market, you can pick up a can of orange juice. But be careful! If you defrost it and try to drink the orange juice straight from the can, you are in for a shock. The orange juice is thick and strong. That's because the can contains orange juice concentrate. Most of the water has been removed from the juice. The juice is too **concentrated** to drink. To prepare it for drinking, you have to put the water back into the juice. Then it is ready to drink.

Orange juice straight out of the orange is a solution. Water is the solvent. There are probably hundreds of solutes in orange juice, including sugars, vitamins, acids, salts, gases, and starches. The particles of all the solutes are evenly distributed among the water particles.

At the orange juice factory, the orange juice is heated to drive off some of the water. As water particles leave the solution, the solute particles (sugar, vitamins, and so on) are still evenly distributed in the solute, but there is less solvent. As a result, the solute particles get closer together, because there are fewer water particles between them. Here's how that works.

Look at the pot of orange juice. There are 100 water particles and 25 orange juice particles.

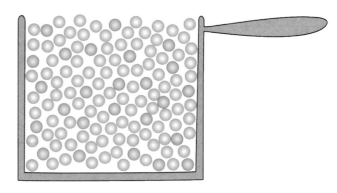

A representation of orange juice in a pot

The pot of juice is gently heated. Solvent (water) particles begin to evaporate.

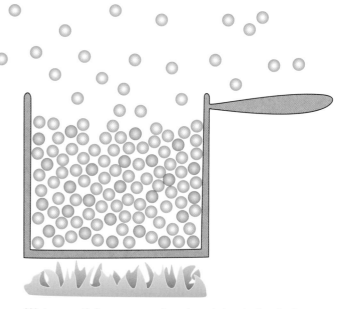

Water particles evaporate when juice is heated.

When 75 water particles have evaporated, the orange juice looks like this.

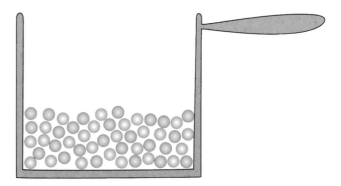

The pot of orange juice after much of the water has evaporated

54

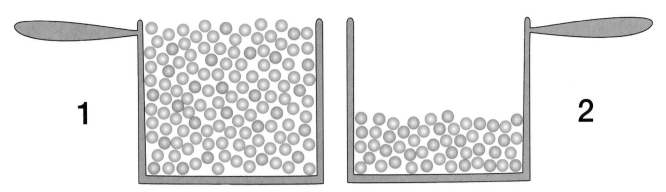

A comparison of a pot of orange juice before (1) and after (2) heating

Concentrated Solutions

Look at the two pots. What's the same and what's different?

Both pots contain the same solvent and solutes: water particles and orange juice particles. And both pots have the same number of orange juice particles. (During heating, only water particles evaporated.)

The important difference is the amount of solvent. The fresh juice in pot 1 had 100 water particles. The evaporated juice in pot 2 has only 25 water particles.

The orange juice in pot 2 is *more concentrated*. That means 100 milliliters (mL) of concentrated solution in pot 2 contains more solute particles than 100 mL of solution in pot 1. The juice in pot 1 is less concentrated.

You can also think of **concentration** as the **ratio** of solvent particles to solute particles. There are four times as many solvent

particles as solute particles in pot 1. The ratio of solvent to solute is four to one. In math, that is written 4:1.

In pot 2, there are 25 solvent particles and 25 solute particles. The ratio of solvent to solute is 1:1. Because there is only one solvent particle for each solute particle in pot 2, that solution is more concentrated than the solution in pot 1. Pot 1 has four solvent particles for each solute particle.

Dilute Solutions

Solutions that are not concentrated are called **dilute** solutions. Dilute solutions have relatively few solute particles "swimming around" among a lot of solvent particles.

After you add water to the concentrated orange juice, it tastes just right. Sweet and delicious. If the juice is not cold enough for your taste, you know how to cool it down. Pour the juice over some ice cubes.

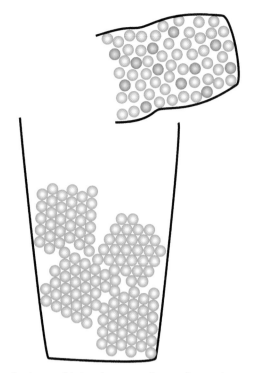

A glass of juice is poured over ice cubes.

The juice will flow down around the ice cubes. Energy will start to flow from the warm juice to the cold ice. The juice will start to cool.

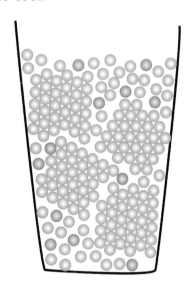

The juice flows around the ice. The juice gets cool and the ice melts.

But there is a catch. When heat transfers to ice, the ice melts. And when ice melts, it changes from solid to liquid. Liquid water

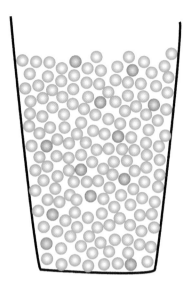

When the ice melts, the orange juice particles are distributed in a lot more water particles.

is the solvent in orange juice. As the ice melts, it dilutes the juice. By the time the ice has all melted, the juice is no longer as tasty as it was.

If you drop the little glass into the big glass of dilute juice, you can compare equal volumes of juice. Which juice sample has more juice particles? Which sample is more concentrated?

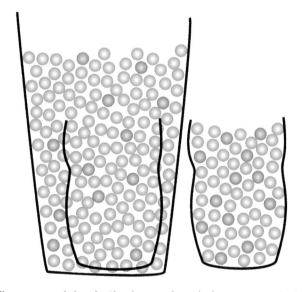

The orange juice in the large glass is less concentrated than it was before it was poured over ice.

General Rules of Concentration

There are two ways to control the concentration of a solution. You can change the amount of solvent, or the amount of solute. The general rules of solutes are these.

- The more solute you use, the more concentrated the solution.

- The less solute you use, the more dilute the solution.

- A solution containing the maximum amount of solute is a **saturated** solution.

The general rules of solvents are these.

- The more solvent you use, the more dilute the solution.

- The less solvent you use, the more concentrated the solution.

Comparing Concentrations

If you have two different solutions made with the same solvent and solute, you can often use a balance to figure out which one is more concentrated. Here's how.

- Measure equal volumes of solution A and solution B.

- Weigh them.

- The sample that weighs more is more concentrated. It's just that simple.

How does that work? Particles have mass. Each solute particle, such as NaCl, has more mass than each solvent particle, such as H_2O. When solute particles go into solution, they take up space. They push some solvent particles out of the way. When you take a sample of a concentrated solution, there will be more heavy solute particles in it than in a dilute solution sample of the same size. So when you put the samples on a balance, the heavier one has more solute particles. More solute particles make the sample more concentrated.

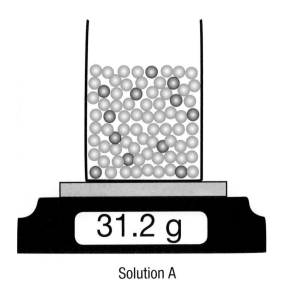

Solution A

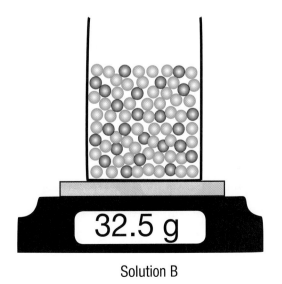

Solution B

Parts per Thousand

Did you ever get a mouth full of seawater? How did it taste? Salty. Very salty. If you evaporate a kilogram of seawater (a little less than a liter), you will be left with about 35 grams (g) of salt. That's about the amount of salt that will sit in the palm of your hand.

A kilogram is 1,000 g. Of those 1,000 g of seawater, 35 g are salt. If you think of grams as "parts," then 35 parts are salt. Concentration of solutes is often reported in parts per thousand (ppt). Thus the salinity (saltiness) of the sea is 35 ppt.

You can certainly taste the salt in seawater. But can you taste the salt in tap water? There is salt in tap water, but the concentration is so low that your taste buds can't identify it. You can do an experiment, however, to discover how concentrated a salt solution has to be in order to taste it. Here's how to set it up.

Dissolve 3.5 g of salt in enough water to make 100 g of solution. That's the concentration of seawater. (Note that 35 ppt is the same as 3.5 parts per hundred.) Transfer exactly 10 g of the "seawater" to a new cup (cup B) and add plain water to it until it has a mass of 100 g.

Now what do you have? The 10 g of seawater transferred to cup B has one-tenth of the salt from the original solution, or 0.35 g. When you dilute it to 100 g, you have 0.35 parts per hundred, or 3.5 ppt. Can you taste the salt? You bet.

Dilute the solution again to get a solution that is 0.35 ppt. Another dilution gives you 0.035 ppt. Can you still taste the salt?

The solution is pretty dilute now. It is now appropriate to shift from parts per thousand to parts per million (ppm). Note that 0.035 ppt is equal to 35 ppm.

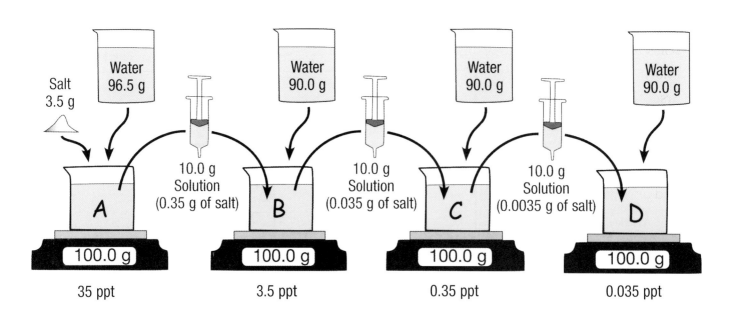

Concentration of Substances in the Environment

Seawater has a high concentration of chlorine because of the salts. Every liter (1,000 mL) of seawater contains 19 grams of chlorine particles. We can refer to this concentration of chlorine as 19 parts per thousand, meaning 19 grams of chlorine per 1,000 mL of water. If we look at 1,000 liters of seawater (one million mL), it will contain 1,000 times as many grams of chlorine. We can say the chlorine has a concentration of 19,000 parts per million (ppm). This system allows us to compare the concentration of other elements in seawater.

In seawater, the element magnesium is pretty concentrated at 1,290 ppm. So are sulfur (904 ppm), calcium (411 ppm), and potassium (392 ppm). Elements in lower concentrations include carbon (28 ppm), nitrogen (16 ppm), and silicon (3 ppm).

Most of the remaining elements are found in seawater, too, but in very low concentrations. For instance, phosphorus (0.09 ppm) is less than 1 ppm. Iron and arsenic (both 0.003 ppm) are significantly less than 1 ppm. These concentrations would be better reported in parts per billion: phosphorus (90 ppb), iron (3 ppb), and arsenic (3 ppb).

Mercury

Mercury is element number 80 on the periodic table. It occurs naturally in Earth's crust and in the ocean. Mercury is a toxic substance for most organisms. In humans, it can cause loss of coordination, irregular heart rate, confusion, muscle and joint pain, and in extreme cases, death. In other organisms, it can interfere with reproduction and cause mutations.

Most of the time mercury is not a problem for living organisms. This is because the concentration is low. In seawater, mercury concentration is 0.2 ppb. That means there are only two mercury particles in every 10 billion particles of seawater.

Problems occur when the concentration of mercury increases. The concentration of mercury in drinking water must not be greater than 2 ppb. At this concentration, the amount of mercury a person would collect in a lifetime would not be dangerous.

But mercury can enter our bodies in other ways, mostly in the food we eat. Large fish that live a long time, like swordfish and sharks, have the most mercury (1 ppm) in their tissues. This is 500 times more concentrated than the acceptable mercury limit in drinking water. Other fish high on the food chain, like tuna, groupers, and sea bass, also have significant mercury concentrations. Small fish and shellfish that grow fast don't live long enough to concentrate a lot of mercury. The lowest concentrations are in sardines, herring, shrimp, oysters, and clams.

The Frog Story

Scientists have noted a significant decline in the worldwide frog and toad populations in the last decade. The causes of the declines are difficult to pin down. There is evidence, however, that toxic chemicals may play a role.

Tyrone B. Hayes (1965–) is a biology professor at the University of California at Berkeley. He has studied frogs for most of his life. The first frogs he saw were in a swamp near his home in South Carolina. Now Hayes studies frogs across the planet, in Africa and North America.

Hayes found something strange going on with some of the frogs he studied. Frogs living in the wild were experiencing sex changes. The "male" frogs were producing eggs. The males were changing into females.

When Hayes and his team analyzed the frogs' environment, they found traces of a common **herbicide** (*herb* = plant; *cide* = kill) in the water. The concentration was only 0.1 ppb. That's one particle in 10 billion. But more tests in the lab showed that the herbicide was causing the changes.

The herbicide does not kill the frogs. But it does affect the population. How? By altering the reproductive processes of the frogs. The substance in the frogs' environment, even in very low concentration, makes it impossible for the frogs to produce offspring. The pesticide kills the next generation, not the current one.

Tyrone B. Hayes

There are about 20,000 different pesticides in use in the United States today. Each one is designed to kill unwanted organisms. But the pesticides don't all stay where they are applied. Particles of the substances are carried by wind and water to other locations. In other environments, the pesticides can have unintended results. This problem is growing all over the world, requiring government regulation and careful use of substances that might have an impact on natural environments.

Carbon Dioxide in the Atmosphere

By now you are aware of **global warming**. The concept is simple: Earth is heating up. What's not so simple is why the heat is on. One scientific fact, however, is known. The concentration of carbon dioxide (CO_2) in the atmosphere is increasing. The cause of the increase is recognized as well. CO_2 is a product of **combustion**. We are burning more fossil fuel than ever before. We are pouring CO_2 into the air 24 hours a day. The problem is made worse by the fact that the forests of the world are being destroyed. Forest plants remove CO_2 during photosynthesis. More CO_2 is entering the air and less is being taken out. The math is pretty simple.

What is the concentration of CO_2 in the air? It varies. Probably the best place to measure the CO_2 concentration is the monitoring station on top of Mauna Loa, a 4,170-meter (13,680') volcano on the big island of Hawaii. This station is out in the middle of the Pacific Ocean, far from major sources of CO_2 and forests. The measurements are a good indicator of the CO_2 concentration worldwide.

When CO_2 monitoring started in 1959, the average concentration was 316 ppm. In 2004, the concentration had risen to 377 ppm. In 45 years, the concentration has increased 61 ppm, or 19%. The graph shows the rapid incline in CO_2 concentration.

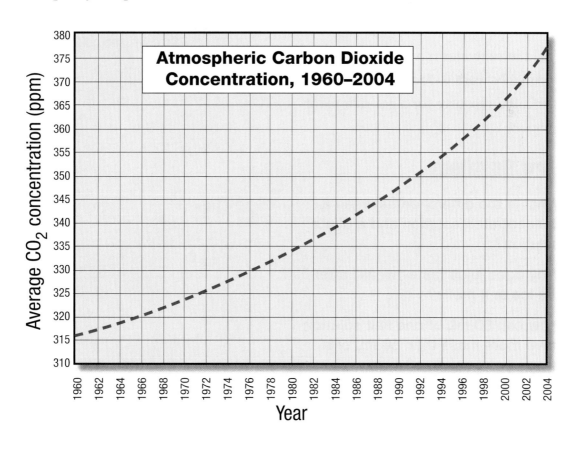

Atmospheric Carbon Dioxide Concentration, 1960–2004

Gold in the Ocean

Here's something to think about. The ocean is probably the most complex solution in the world. Just about everything you can think of is dissolved in there. One element in seawater is gold. The concentration of gold is about 0.00001 ppm. The question is how to get it out. You need something like a gold magnet.

If you did invent a gold magnet, could you get rich? Maybe. If you do the math, it looks like you could extract 200 g of gold (not quite half a pound) from every 180 billion liters of seawater. How much water is that? About the amount it takes to fill 75,000 Olympic-sized swimming pools.

The gold in seawater is a little dilute. You probably want to focus on a more concentrated resource for a get-rich plan!

Gold coins and gold bars

Review Questions

1. What is the difference between a concentrated solution and a dilute solution?

2. Why does juice taste "weak" after the ice in it melts?

3. How can 50 mL of one salt solution have more mass than 50 mL of a second salt solution?

4. What is the maximum concentration of mercury allowed in drinking water in the United States?

How Do Atoms Rearrange?

When you blow air through a straw into a cup of limewater, something happens. After a couple of breaths, the limewater is just noticeably hazy. After a couple more breaths, the limewater is definitely cloudy. After a dozen breaths, the cup of limewater is as white as milk.

What turns the limewater white? Plain air bubbled through limewater does not turn limewater white. Something in your breath does it. Carbon dioxide. It is the carbon dioxide in your breath that turns the limewater white.

That's not exactly right. Carbon dioxide isn't white, so it doesn't turn the limewater white. What happens is carbon dioxide mixes with something in the limewater to form a white substance. The white substance is not carbon dioxide or limewater. It is a new white substance.

What's happening is a **chemical reaction**. During a chemical reaction, starting substances change into new substances. Starting substances are called **reactants**. New substances are called **products**.

Figuring Out New Products

Carbon dioxide is one of the reactants. Limewater is the common name for a solution of calcium hydroxide and water. Calcium hydroxide is the other reactant. You can set up a **chemical equation** to help figure out what the white product might be. The **chemical formula** for carbon dioxide is CO_2. The formula for calcium hydroxide is $Ca(OH)_2$. The reactant side of the equation looks like this.

$$CO_2 + Ca(OH)_2 \rightarrow$$

This is how you read the equation: "One particle of carbon dioxide and one particle of calcium hydroxide react to yield…"

What could the products be? A set of **atom** tiles might help. Make representations of a carbon dioxide particle and a particle of calcium hydroxide. You can then move the atoms around to make products. An atom-tile representation of the equation looks like this.

O C O + H O Ca O H →

Carbon dioxide   plus   Calcium hydroxide   yields

The white substance that formed in the reaction did not dissolve in water. That's a clue. You need to see what insoluble substance you can make by rearranging the atoms in the reactants. Remember, calcium carbonate ($CaCO_3$) doesn't dissolve in water. The atoms needed to make a calcium carbonate particle are in the reactants. You can move those atoms to the product side of the equation.

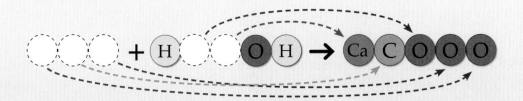

What's left on the reactant side of the equation? Two hydrogen atoms and one oxygen atom. These combine to form a particle of water on the product side.

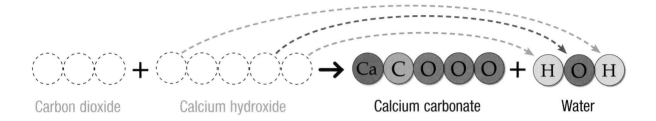

Carbon dioxide   Calcium hydroxide   Calcium carbonate   Water

Notice that after the reaction has happened, the reactants are gone. They no longer exist. All the atoms that were in the reactants are now in the products. And the number of atoms on the product side is exactly the same as the number of atoms that started on the reactant side. During reactions, no atoms are destroyed or created. Matter is conserved. That means the number of atoms must be the same on both sides of the equation. This is a hard-and-fast rule.

Written with chemical formulas, the limewater equation looks like this. The down arrow by the formula for calcium carbonate means the substance appears as a **precipitate**.

$$CO_2 + Ca(OH)_2 \rightarrow CaCO_3\downarrow + H_2O$$

Other Reactions

Did you ever see a blimp slowly sailing along overhead? It is an interesting vehicle. It stays aloft by being lighter than air. It actually floats in air in a way that is similar to a cork floating in water. To make the blimp lighter than air, it is filled with helium, element number 2. The blimp is a huge motorized helium balloon.

The hydrogen-filled *Hindenburg* burned completely in 37 seconds.

During the 1930s, lighter-than-air craft were developed for transportation. One of the largest such craft ever built was the *Hindenburg,* a 245-meter-long (800′) hydrogen-filled monster. On May 6, 1937, the largest structure ever to fly caught fire. In 37 seconds, the entire massive structure was destroyed.

The hydrogen burned. Burning is a reaction called combustion. The reactants were the hydrogen gas (H_2) in the *Hindenburg* and the oxygen (O_2) in the air. You can use atom tiles to see what products formed.

$$\text{H H} + \text{O O} \rightarrow \text{H O H}$$

Water! When hydrogen burns, the single product is water. But look, something is wrong with the equation. There is one oxygen atom left over. How can this be fixed? The solution is to react *two* hydrogen particles with one oxygen particle to form two particles of water. Here is the balanced equation.

$$\text{H H} + \text{H H} + \text{O O} \rightarrow \text{H O H} + \text{H O H}$$

The field of modern chemistry grew out of experimental work with substances, called **alchemy**. One of the goals of the early alchemists was to turn common metals into gold. They never succeeded.

But changing one metal into another with a simple reaction is possible. If you have a solution of copper chloride ($CuCl_2$) and drop in a piece of aluminum foil (Al), a reaction takes place.

Aluminum metal and copper chloride react to form aluminum chloride and copper metal.

When the reaction is complete, the silvery aluminum is gone. In its place is a reddish brown precipitate of copper (Cu) metal and a solution of aluminum chloride ($AlCl_3$). Look at the reaction equation.

$$CuCl_2 + Al \rightarrow AlCl_3 + Cu$$

There is a problem. Do you see it? The chlorine atoms are not balanced, so it is not possible for one particle of each reactant to react. If you start with two $CuCl_2$ particles, then you have four chlorine atoms. That's too many. To balance the equation, you need to start with three $CuCl_2$ particles and two Al particles. The balanced equation looks like this.

$$3CuCl_2 + 2Al \rightarrow 2AlCl_3 + 3Cu$$

Reaction Summary

The particles of all substances are made of atoms. The kind, number, and arrangement of atoms determines the kind of substance.

New substances are created during chemical reactions.

During reactions, atoms rearrange. The atoms in the particles of the reactant substances rearrange to form the particles of the products.

Reactions can be described with equations. Substances in equations can be recorded with atom representations or chemical formulas.

Reaction equations must be balanced. The number of atoms of each kind must be equal on both sides of the equation. Balance is achieved by changing the number of particles reacting.

Matter is conserved. Particles of substances are created and destroyed during chemical reactions. Atoms are *not* created or destroyed during chemical reactions. Atoms rearrange to create new particles of substances.

Review Questions

1. What is destroyed and what is created during chemical reactions?

2. What are reactants and products? Write a reaction equation and label the reactants and products.

3. Write the equation for the reaction between hydrogen and oxygen. Use chemical formulas for the substances.

4. Methane (CH_4) is the main gas in natural gas. The products that form when methane burns are carbon dioxide and water. Write a balanced equation showing the combustion reaction when methane and oxygen react.

Antoine-Laurent Lavoisier
The Father of Modern Chemistry

In 1794, during the French Revolution, a man was executed. This man was a famous chemist who helped revolutionize the study of chemistry. His experiments changed the way chemists understood fire. His work stressed the importance of careful measurement in experiments. He created a new language of chemistry and defended the idea that matter is not lost or gained during chemical reactions. Why was Antoine-Laurent Lavoisier's head cut off? Was it because of these ideas? Perhaps looking at his life and his work will help us understand.

Lavoisier (1743–1794) was born when King Louis XV ruled France. Because his family was wealthy, young Lavoisier was able to get an education. As a young man, he practiced law and owned a tax-collecting agency. Lavoisier was a prominent citizen in his community.

But it is Lavoisier's second life as a scientist for which he is remembered. Early in the morning and late at night, he studied mathematics, geology, physics, biology, and chemistry. His work in geology was so impressive that he was elected to the French Academy of Sciences when he was 25 years old.

Portrait of Lavoisier in his laboratory

Debunking Phlogiston

Lavoisier's interest in science turned more and more toward chemistry. One of many things he studied was combustion, the chemical reaction we usually call burning. During the 1600s and 1700s, chemists thought that all flammable substances contained an odorless, colorless substance called phlogiston. It was generally thought that when a substance burned, it gave up this phlogiston, and the material left behind weighed less than it did before.

Lavoisier thought the theory of phlogiston was nonsense. He called it a "fatal error to chemistry." He set out to disprove the phlogiston theory.

In 1772, Lavoisier performed a series of experiments based on the work of another French chemist, Louis-Bernard Guyton de Morveau (1737–1816). He kept very detailed records, carefully measuring the weights of materials before and after he burned them. His results showed that substances such as sulfur and phosphorus actually gained weight when burned.

In 1774, Lavoisier met the English chemist Joseph Priestley (1733–1804). Priestley told him that he had discovered a "new kind of air" by burning mercuric oxide (HgO). Priestley tested animals in a closed container with this new air and found that they lived longer than animals in closed containers with regular air.

Intrigued, Lavoisier repeated Priestley's mercuric oxide experiments. But, unlike Priestley, he very systematically weighed and measured all parts of his experiments. He collected the gas given off and found that a candle thrust into the gas burned brighter than he could have imagined. Could Priestley's new air somehow be involved in the burning of other substances? Could it also explain why the substances he had burned before gained weight?

To find answers to his questions, Lavoisier devised a new piece of equipment. It was a kind of carefully crafted air-tight reaction chamber and oven. He put liquid mercury (Hg) into the reaction chamber, put the chamber in the oven, and heated it for 12 days.

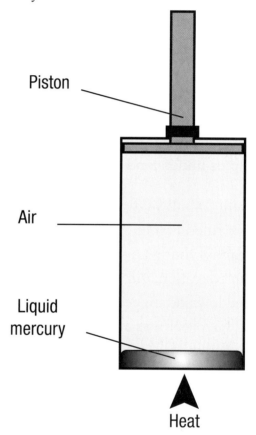

Three interesting things happened. The mercury turned into a reddish solid substance. The volume of air in the container decreased from 50 to 42 cubic inches. And the new solid substance weighed more than the mercury!

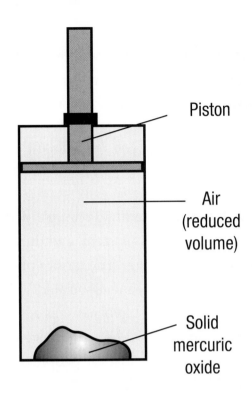

Piston

Air (reduced volume)

Solid mercuric oxide

Lavoisier concluded that the missing 8 cubic inches of air had combined with the shiny liquid mercury to produce the dull red mercuric oxide. The additional weight of the mercuric oxide was exactly equal to the weight of the "missing" 8 cubic inches of air.

He then heated the solid mercuric oxide to a higher temperature. It turned back into liquid mercury and gave off the exact amount of gas that had been lost before.

This change was proof to Lavoisier that combustion does not burn phlogiston out of a substance, but rather combines the substance with part of the air surrounding the substance. In 1785, he said that his ideas should replace the idea of phlogiston. Lavoisier later named this ingredient in air *oxygen,* a name that we still use.

Some important chemists, including Priestley, who first isolated oxygen and called it "new air," disagreed with Lavoisier's conclusion. They remained solid supporters of the phlogiston theory. But, in the end, the precisely measured results of Lavoisier's experiments convinced the new generation of chemists that oxygen explained combustion.

Other Major Contributions

What impressed Lavoisier's supporters more than anything else was his careful attention to quantitative detail. He spent a lot of his own money (he was wealthy) designing and building instruments, including balances as precise as those used today. With these instruments, he was able to establish the law of conservation of mass, which states that matter is neither gained nor lost during a chemical reaction.

Lavoisier was also quite the showman, making sure that his equipment was not only precise but flashy. One piece he developed, the gasometer, was used to measure volumes of gases. It stood about 6 feet tall, was made of gleaming brass, and was one of the most expensive pieces of chemistry apparatus anyone had ever seen. It cost more than $250,000 in today's dollars. Although its results were no better than anyone else's, its sheer expense and beauty made people begin to think of chemistry as serious science.

Chemistry, a serious science? Many people at that time still thought of chemistry as magic. But Lavoisier's Chemical Revolution was changing that. In addition to his experiments, he developed the system of chemical names that still exists today. This new vocabulary enabled chemists to communicate effectively and established chemistry as an independent, respected science.

To top it all off, he wrote what is now considered to be the first textbook of chemistry, *The Elements of Chemistry,* in 1789. In it, he introduced the idea of "elements," substances that could not be broken down into simpler substances. He included heat and light as substances, though they are now known to be energy.

Lavoisier was wrong about one major idea. When presented with the idea of atoms, tiny, unchangeable, defining units of an element, he dismissed them as impossible.

Death at Such a Young Age

Were these new ideas threatening to established scientific thought? Did Lavoisier deserve to lose his head for his ideas about chemistry? Actually, no. Science had been experiencing a revolution in thought called the Enlightenment. The intellectual world was ready to accept arguments based on experimentation and logical thought. Lavoisier and the chemists of the Chemical Revolution used those techniques. But it all came tumbling down for Lavoisier when the French Revolution broke out in 1789, the same year he published *The Elements of Chemistry.*

The French peasants had had enough of the monarchy and class system, and they rebelled against the king. The targets of their fury included anyone associated with the government. Lavoisier was a tax collector, and for that he was arrested, convicted of crimes against the people of France, and beheaded.

Review Questions

1. Why did mercuric oxide in Lavoisier's reaction chamber weigh more than the mercury metal?

2. Why was there less air in Lavoisier's reaction chamber after he heated the mercury for 12 days?

3. What are some of the reasons Lavoisier is considered to be the father of modern chemistry?

Organic Compounds

*W*hen a car pulls into a gas station, the driver is faced with a decision. Which grade of gas should I buy?

Regular is the lowest grade. If you check the pump, you may see that it is labeled 87 **octane**. More expensive grades have higher octane ratings, maybe as high as 92 or 93 octane. Motorists who choose economical cars reach for the regular. Those with high-performance cars must use the high-octane fuels.

What is octane? It is a substance. Its chemical formula is C_8H_{18}. Octane is the main ingredient in gasoline. Gasoline rated at 87 octane is 87% octane.

A gas pump offers three grades of gasoline: 87 octane, 89 octane, and 92 octane.

Under normal conditions octane is a liquid. If it is ignited with a match, it will react with oxygen in air and burn with a yellow flame. But if it is sprayed into a fine mist and ignited with a spark, it will explode. An explosion is a reaction that occurs extremely rapidly, and gives off a lot of energy in the forms of light and heat. That's what happens inside the cylinders of a car engine to make a car go.

The octane reaction produces two products, carbon dioxide and water. Here is the equation. Is it balanced?

$$2C_8H_{18} + 25O_2 \rightarrow 16CO_2 + 18H_2O$$

When the reaction is shown with atom representations, you can see the large number of particles involved in this important reaction.

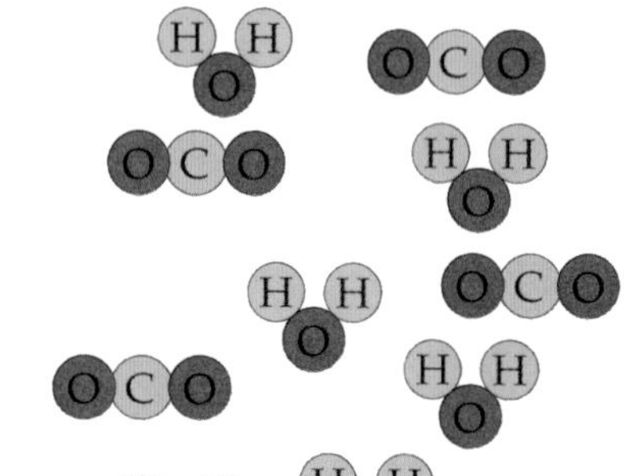

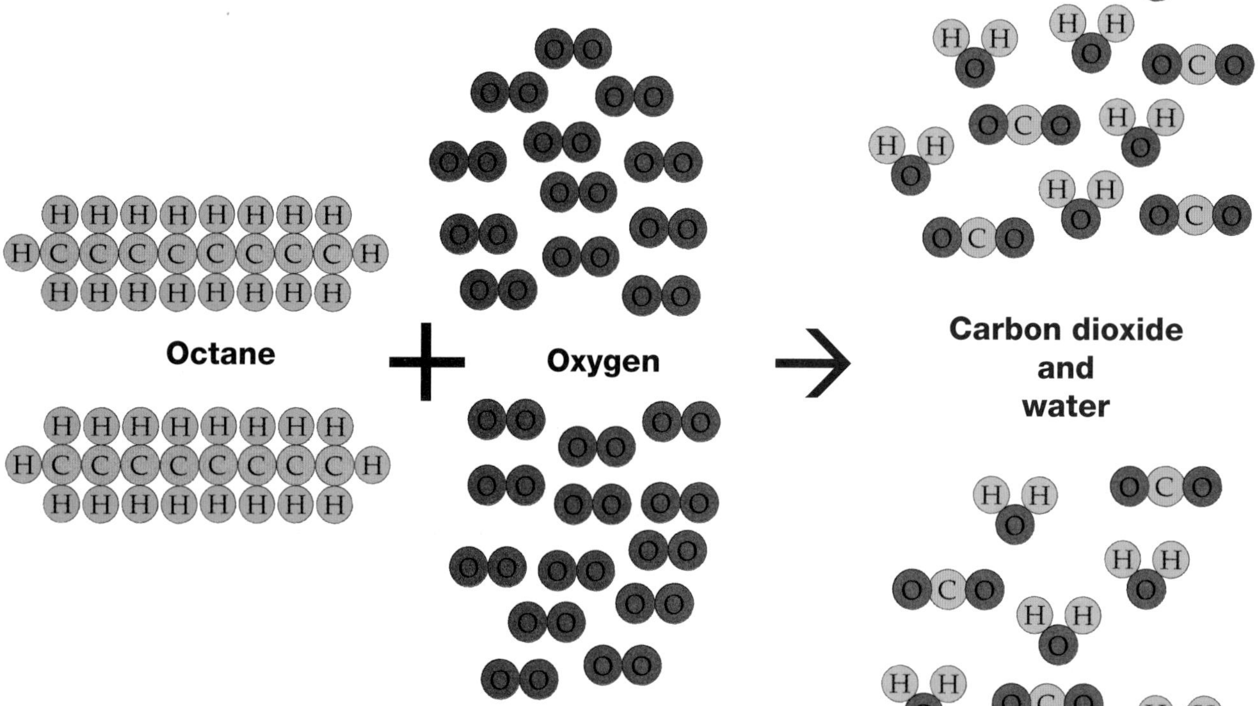

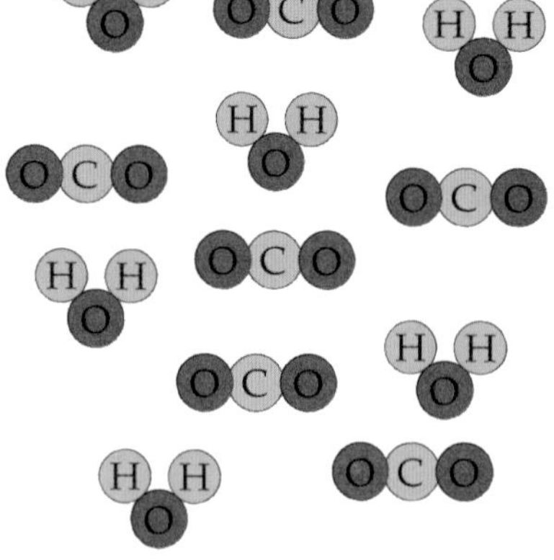

Octane

+

Oxygen

→

Carbon dioxide and water

Hydrocarbons

Octane is one member of a large family of substances called **hydrocarbons**. All hydrocarbons are made from only two elements, hydrogen and carbon. Naturally occurring sources of hydrocarbons include natural gas, petroleum oil, and, to some extent, coal. Hydrocarbon fuels are burned to produce heat, the energy that makes things happen.

One of the simplest hydrocarbons is methane (CH_4). Methane is a gas at standard temperature and pressure. It is the main gas in natural gas, which is used in homes and businesses for cooking and heating. Propane (C_3H_8) is a hydrocarbon with a slightly larger **molecule**. Propane is also gas at standard temperature and pressure. But propane under pressure changes into liquid. That property makes propane easy to store and transport.

People who live in remote areas, or who travel in large recreational vehicles, use propane for fuel. The fuel is kept in pressurized bottles or tanks. When the valve on the storage tank is opened, the propane turns back into gas. The gas is used to cook dinner.

The table below shows the chemical formulas for the hydrocarbons from methane to octane.

Portable propane tanks

A large propane tank stores fuel behind a home.

Simple Hydrocarbons		
Substance name	Number of carbon atoms	Chemical formula
Methane	1	CH_4
Ethane	2	C_2H_6
Propane	3	C_3H_8
Butane	4	C_4H_{10}
Pentane	5	C_5H_{12}
Hexane	6	C_6H_{14}
Heptane	7	C_7H_{16}
Octane	8	C_8H_{18}

Molecules of Life

Hydrocarbons are part of a larger category of substances called **organic compounds**. Organic compounds are the substances of life. They are made by living organisms.

The most important element in organic chemistry is carbon. The carbon atom has the unique ability to bond with other carbon atoms to make long chains and rings. The huge number of ways carbon atoms can combine results in an endless variety of molecules. All those millions of different carbon-based organic molecules create the diversity of life on Earth.

The importance of carbon chaining can't be overstated. Many scientists think that if there is life in other places in the universe, it will be carbon based. No other atom has the ability to produce the complexity needed for life.

Getting Your Carbon

Did you have your carbon today? You need a constant supply to stay healthy and fit. Don't worry, you are getting plenty of carbon. Here's how.

You have probably heard the story of photosynthesis. Plants and algae use water, light from the Sun, and carbon dioxide to make food. Well, the food that plants make is sugar, which is a carbon chain with hydrogen and oxygen atoms bonded to it. The carbon comes to the plant one carbon atom at a time in the form of carbon dioxide gas. The CO_2 molecules are taken apart and reassembled into sugar in a series of complex reactions.

The plant uses the sugar for energy. The end products of that process are energy, water, and CO_2. But there is more. Some of the sugar produced by photosynthesis is turned into other organic substances. Those substances include fats, proteins, and starches. These substances make all the other plant structures, like stems, leaves, flowers, seeds, and so on.

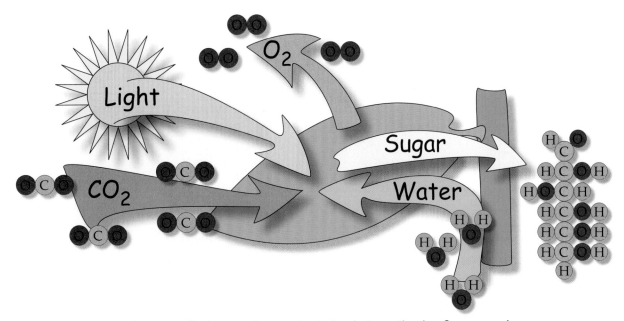

Water and carbon dioxide are the reactants in photosynthesis. Oxygen and sugar are the products. The reaction also uses energy from sunlight.

It doesn't matter how long you sit out in the sunshine, you can't make sugar out of CO_2. So how do you get your carbon? You eat carbon-rich organic material. That means plants, plant parts, and animals. That's where your carbon is. Your body breaks down the carbon-rich substances during digestion and uses them to make the molecules you need to survive. You are a very advanced chemical factory, producing thousands of different carbon-based substances all the time.

Some of the carbon-based molecules are used for energy. The carbon chains are taken apart and rearranged into new substances in a complex series of reactions. That process produces two end products, water and CO_2. Water can be recycled, but CO_2 is a toxic waste that must be eliminated. How do you do that? Red blood cells carry the CO_2 molecules one by one to the lungs where they are exhaled. And you know how to prove that the CO_2 is returning to the air where it originally came from. Just blow some exhaled air through limewater and...

Dr. Donna Nelson — Chemist

Chemistry is everywhere you look. When you strike a match, the way a cake rises in the oven, or when your bicycle begins to rust if you leave it out in the rain. All of this happens because of chemical reactions—when two or more molecules interact and something happens.

In my work as a physical organic chemist, I study the rate and speed of chemical reactions and then I compare and contrast them. Sometimes I will modify one of the molecules to see if it changes and how quickly the reaction occurs. For example, a fireworks display is a complex series of chemical reactions. The colors in fireworks are made up of many different chemicals combined together to react with heat and energy. The amount of heat and energy applied will change the way the colors of fireworks look. In addition, by varying the amount of the chemicals responsible for a particular color, you can make different shades, so basic red could become maroon or pink.

Just like the many chemical ingredients in fireworks, there are many chemical ingredients that make up things we use every day from medicine to laundry soap. Every single one of these chemicals needs to be tested and re-tested to see their reactions in many different types of

Dr. Donna Nelson

situations. The people that develop new things like soap and medicine are called synthetic chemists. In my job as an organic chemist, I provide these synthetic chemists with important information about their ingredients. Thus, my work contributes to making sure that all of the ingredients of a new product work well together.

The work that I do is very exciting and some days it feels like a big brain teaser. As a kid I always loved brain teasers. Although this might have been the first clue that I was going to be a scientist, I always thought I was going to be the town doctor just like my father.

I grew up in the small town of Eufaula, Oklahoma, and at times, my father was the only doctor around. I remember once when we came home from a movie we saw people sitting on our lawn waiting for us. There had been an accident and because the closest hospital was 30 miles away, the injured people were taken to our house.

My father found the life as a small town doctor challenging. He never said I couldn't do it, but he and my mother supported my other interests like math and science. With my parents' encouragement, I went on to the University of Oklahoma where I decided to major in chemistry.

In Eufaula, almost everybody is at least some part Cherokee, Chickasaw, or Creek, so going to a big university, where there were barely any Native Americans, was a big shock. It took time to adjust to being one of very few women and the only Native American in my department. I survived by being quiet in school, but very persistent.

When I decided to go to graduate school at the University of Texas at Austin, people laughed at me. They said, "You go to graduate school?!" I learned to ignore their comments and have confidence in myself. I graduated with my Ph.D. in chemistry in 1980. After doing research at Purdue University, I went back to the University of Oklahoma to become a professor of chemistry.

I am still the only Native American and one of only three women in my whole department, but I am no longer quiet! I work very hard to educate people about the status of women and minorities in the sciences at universities around the country. I have learned persistence and confidence are the keys to success, and that speaking the truth, even if it is hard for people to hear, helps everyone.

New Technologies

I n 1907, Ernest Rutherford (1871–1937) discovered the structure of the gold atom. His experiments suggested a solid, positively charged **nucleus** with negatively charged **electrons** orbiting it. That was a breakthrough.

Using Rutherford's ideas, scientists were able to figure out the structure of all the atoms. With models for structures of the atoms, scientists could predict how atoms might combine to form molecules of substances. They predicted that atoms would arrange in **well-ordered arrays**. Well-ordered arrays are repeating patterns, like marbles in a box or oranges carefully stacked at the market.

For years it was impossible to find out if the predictions were right. Atoms were way too small to see, even with the world's most powerful microscopes.

That changed in 1981 when a new kind of microscope was invented. The **scanning tunneling microscope (STM)** could create images of individual atoms! For the first time, scientists could see that atoms did arrange themselves in orderly arrays, just as they had predicted. Each white dot in the image to the right is one atom of silicon. The silicon atoms form an interesting pattern of circles.

A researcher with a scanning tunneling microscope (STM)

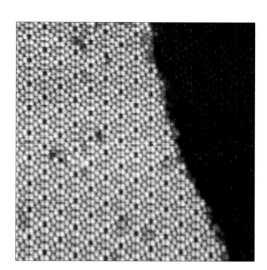

Silicon atoms in a silicon crystal

The Search for New Elements

By 1930, all 90 naturally occurring elements had been discovered and described. The periodic table went from element number 1, hydrogen, to element number 92, uranium. Elements 43 and 61 had not been discovered. They just didn't seem to exist on Earth. And no elements existed beyond number 92, uranium. Element number is also called **atomic number**.

Scientists knew that if they could put one more **proton** in the nucleus of a uranium atom, the result would be a new element, element number 93. But how could they do that?

In 1931, a young professor at the University of California at Berkeley, Ernest O. Lawrence (1901–1958), invented an instrument called the **cyclotron**. The cyclotron was the tool that made it possible to create new elements. It used electricity to accelerate protons and other atomic particles. The particles would move so fast that they would crash into the nucleus of uranium and stick there. In 1940, the cyclotron was used to make element number 93, neptunium.

Glenn Seaborg (left) and Albert Ghiorso

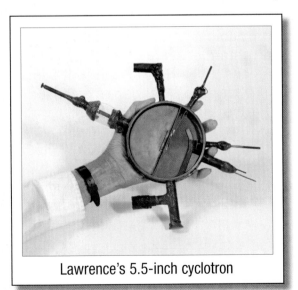
Lawrence's 5.5-inch cyclotron

Between 1940 and 1974, elements 93 through 106 were created in the labs at Berkeley. The work was done by teams, but two scientists provided leadership. They were Glenn Seaborg (1912–1999) and Albert Ghiorso (1915–). Not only did they decide which atoms to collide to make new elements, they designed bigger and bigger cyclotrons. By the 1970s, the cyclotrons were huge buildings.

One of several particle accelerators at Berkeley

In the 1980s and 1990s, elements 107–112 were discovered. The work was done in Germany at the Heavy Ion Research Lab. The creation of elements 113–116 has been reported. The work was done in Russia by international teams of scientists. The discoveries have not yet been confirmed. It may be years before the experiments are repeated by other scientists to prove that the elements really exist.

Recently scientists at the Berkeley lab announced the discovery of elements 117 and 118. After failing to confirm the discovery, the scientists withdrew their claim. But pay attention to the science news. In a few years, you might hear about the discovery of element 120. Do you think there is a limit to the size of atoms scientists can make?

James Harris

(1932–)

In 1969, James Harris was a member of the team of scientists who discovered elements 104 and 105. He was the first African American scientist to discover new elements. Harris designed the part of the experimental system called the target. The target is necessary to identify the products of the high-energy collisions that produce new elements.

Lawrence Hall of Science, dedicated in 1968

Ernest O. Lawrence

Ernest O. Lawrence

The discovery of 26 transuranium (beyond uranium) elements has been possible because of the cyclotron invented by Lawrence. His tiny cyclotron started a whole new branch of science. During his 57 years, he accomplished a lot. He received the Nobel Prize in 1939, making him the first person at the University of California to be awarded the prize. Two research laboratories are named for him.

After his death, Lawrence's family and fellow scientists wanted to build a memorial in his honor. They decided to build a science museum. It is the Lawrence Hall of Science. Every year thousands of families and school classes visit the museum. The Lawrence Hall of Science is an active, ever-changing science experience. It is a fitting memorial to a man who started a science revolution.

Review Questions

1. How do scientists know that atoms and molecules combine in well-ordered arrays?

2. Explain why the periodic table has over 110 elements when only 90 occur naturally on Earth.

Gertrude B. Elion

Have you heard of the disease leukemia? It is a blood cancer. Before 1950, half of the children who got leukemia died in a few months. That was before Gertrude B. Elion (1918–1999) created a molecule that attacked the cancer. The molecule's scientific name is 6-mercaptopurine, but it is usually called 6-MP. Elion made the molecule with a reaction that replaced an oxygen atom with a sulfur atom. The new 6-MP molecule was a breakthrough.

Gertrude Belle Elion

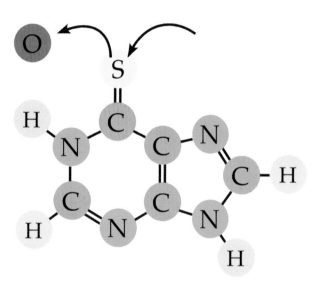

A model of the 6-MP molecule shows where a sulfur atom replaced an oxygen atom.

Elion decided to become a scientist soon after high school graduation. The decision followed a very unhappy period in her life, during which she watched helplessly as her grandfather died of cancer. She resolved to find a cure for cancer.

In 1937, Elion graduated from Hunter College in New York with a degree in chemistry. She applied for graduate studies at many colleges, but was turned down. She went from company to company looking for work in a laboratory, but was told over and over that there were no jobs for women. At that time, women did not work in laboratories.

Discouraged, Elion went to secretarial school. She took a series of temporary jobs in order to save enough money to attend graduate school. At New York University, she was the only woman in her chemistry classes. It was 1942, and the United States had entered World War II. Jobs for women were opening up. Elion got her first lab job doing food-quality testing—inspecting fruit for mold, checking the acidity of pickles, and so on. The work wasn't what she wanted, but she did learn to use a lot of scientific instruments.

In 1944, Elion got her first research position with a drug company called Burroughs Wellcome. She was 26 years old. There she met her mentor, George Hitchings, who guided her development as a researcher. She started doing the medical research she had been aiming for during the previous 10 years.

Elion refers to 1950 as her "Wow" year. That was the year she developed 6-MP and a similar molecule for fighting cancer. She was 32 years old.

For the next 30 years Elion devoted her life to developing new **compounds** to help the human body fight disease. But she was not motivated by the satisfaction of producing a new, useful molecule. She thought about her work differently. She studied carefully the interactions between diseases and cells. She wanted to know why cancer cells were so successful. What were they doing, and what was allowing them to do it? When she figured out what the cancer needed in order to succeed, she attempted to produce a molecule that confused or injured the cancer. She focused her intelligence on interactions at the cellular level.

In 1988, Elion received the Nobel Prize for Medicine. This was a remarkable achievement, because she was neither a medical doctor nor a PhD chemist. She won her acclaim the old-fashioned way, through hard work and determination.

When Elion first joined Burroughs Wellcome in 1944, she intended to stay with the company only as long as she had the opportunity to learn new things. Elion never left. She found challenge after challenge to occupy her mind. Her curiosity led her through the fields of biochemistry, pharmacology, immunology, and virology. Along the way her research confronted not only cancer, but also the challenges of organ transplant rejection; viral infection, including herpes and HIV; and a vexing, painful condition called gout.

FOSS Chemical Interactions Resources

Table of Contents

References

SAFETY PRACTICES

Always follow these safety practices when you do laboratory activities.

1. Follow the safety procedures outlined by your teacher. Ask questions if you're unsure of what to do.

2. Conduct yourself in a responsible manner. No running, pushing, horseplay, or practical jokes are allowed.

3. Know where classroom safety equipment is located and know how to use it.

4. Wear protective eyewear (goggles) when instructed to do so by your teacher.

5. Never eat, drink, or chew gum during laboratory activities.

6. Use care and common sense when working with chemical substances.

   - Never put substances in your mouth. Do not taste any substance unless your teacher specifically tells you to do so.

   - Do not smell unknown substances. If your teacher asks you to smell a substance, wave a hand over the substance to draw the scent toward you.

   - Do not touch your face, mouth, eyes, or another person while working with chemical substances.

   - Wash your hands with soap and warm water immediately after working with chemical substances.

7. If you come in contact with a chemical substance, IMMEDIATELY rinse skin and clothing with water, and THEN inform your teacher.

8. Report all accidents and injuries to your teacher.

9. Use extra care when working around open flame.

   - Pull back and secure long hair.

   - Do not leave a burning candle unattended.

   - Do not reach over an open flame.

   - Do not use bare hands to pick up hot objects.

   - Wear protective eyewear.

10. Cleanup is YOUR responsibility! Handle all used and unused substances as directed by your teacher. Clean up your work area.

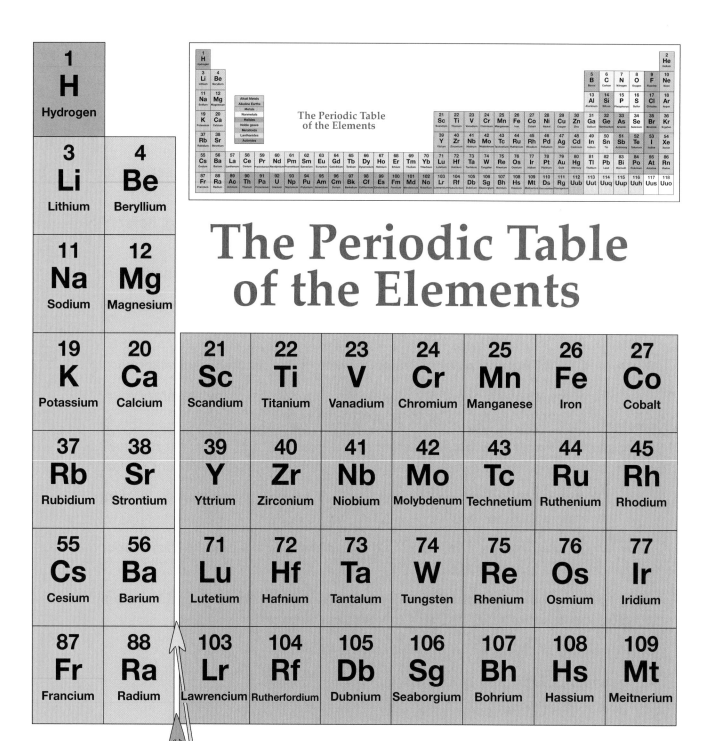

The Periodic Table
of the Elements

Legend

- Alkali Metals
- Alkaline Earths
- Metals
- Nonmetals
- Halides
- Noble gases
- Metalloids
- Lanthanides
- Actinides

Periodic Table

					2 **He** Helium
5 **B** Boron	6 **C** Carbon	7 **N** Nitrogen	8 **O** Oxygen	9 **F** Fluorine	10 **Ne** Neon
13 **Al** Aluminum	14 **Si** Silicon	15 **P** Phosphorus	16 **S** Sulfur	17 **Cl** Chlorine	18 **Ar** Argon

28 **Ni** Nickel	29 **Cu** Copper	30 **Zn** Zinc	31 **Ga** Gallium	32 **Ge** Germanium	33 **As** Arsenic	34 **Se** Selenium	35 **Br** Bromine	36 **Kr** Krypton
46 **Pd** Palladium	47 **Ag** Silver	48 **Cd** Cadmium	49 **In** Indium	50 **Sn** Tin	51 **Sb** Antimony	52 **Te** Tellurium	53 **I** Iodine	54 **Xe** Xenon
78 **Pt** Platinum	79 **Au** Gold	80 **Hg** Mercury	81 **Tl** Thallium	82 **Pb** Lead	83 **Bi** Bismuth	84 **Po** Polonium	85 **At** Astatine	86 **Rn** Radon
110 **Ds** Darmstadtium	111 **Rg** Roentgenium	112 **Uub**	113 **Uut**	114 **Uuq**	115 **Uup**	116 **Uuh**	117 **Uus**	118 **Uuo**

63 **Eu** Europium	64 **Gd** Gadolinium	65 **Tb** Terbium	66 **Dy** Dysprosium	67 **Ho** Holmium	68 **Er** Erbium	69 **Tm** Thulium	70 **Yb** Ytterbium
95 **Am** Americium	96 **Cm** Curium	97 **Bk** Berkelium	98 **Cf** Californium	99 **Es** Einsteinium	100 **Fm** Fermium	101 **Md** Mendelevium	102 **No** Nobelium

The Elements in Alphabetical Order

Element name	Symbol	Atomic number	Date of discovery	Element name	Symbol	Atomic number	Date of discovery
Actinium	Ac	89	1899/1902	Europium	Eu	63	1901
Aluminum	Al	13	1825	Fermium	Fm	100	1953
Americium	Am	95	1944	Fluorine	F	9	1886
Antimony	Sb	51	Prehistoric	Francium	Fr	87	1939
Argon	Ar	18	1894	Gadolinium	Gd	64	1880
Arsenic	As	33	Middle Ages	Gallium	Ga	31	1875
Astatine	At	85	1940	Germanium	Ge	32	1886
Barium	Ba	56	1808	Gold	Au	79	Prehistoric
Berkelium	Bk	97	1949	Hafnium	Hf	72	1923
Beryllium	Be	4	1797	Hassium	Hs	108	1984
Bismuth	Bi	83	1753	Helium	He	2	1868
Bohrium	Bh	107	1981	Holmium	Ho	67	1878
Boron	B	5	1808	Hydrogen	H	1	1766
Bromine	Br	35	1826	Indium	In	49	1863
Cadmium	Cd	48	1817	Iodine	I	53	1811
Calcium	Ca	20	1808	Iridium	Ir	77	1804
Californium	Cf	98	1950	Iron	Fe	26	Prehistoric
Carbon	C	6	Prehistoric	Krypton	Kr	36	1898
Cerium	Ce	58	1803	Lanthanum	La	57	1839
Cesium	Cs	55	1860	Lawrencium	Lr	103	1961
Chlorine	Cl	17	1774	Lead	Pb	82	Prehistoric
Chromium	Cr	24	1797	Lithium	Li	3	1817
Cobalt	Co	27	1739	Lutetium	Lu	71	1907
Copper	Cu	29	Prehistoric	Magnesium	Mg	12	1808
Curium	Cm	96	1944	Manganese	Mn	25	1774
Darmstadtium	Ds	110	1994	Meitnerium	Mt	109	1982
Dubnium	Db	105	1970	Mendelevium	Md	101	1955
Dysprosium	Dy	66	1886	Mercury	Hg	80	Prehistoric
Einsteinium	Es	99	1952	Molybdenum	Mo	42	1778
Erbium	Er	68	1843	Neodymium	Nd	60	1885

Element name	Symbol	Atomic number	Date of discovery
Neon	Ne	10	1898
Neptunium	Np	93	1940
Nickel	Ni	28	1751
Niobium	Nb	41	1801
Nitrogen	N	7	1772
Nobelium	No	102	1958
Osmium	Os	76	1803
Oxygen	O	8	1774
Palladium	Pd	46	1803
Phosphorus	P	15	1669
Platinum	Pt	78	1735/1741
Plutonium	Pu	94	1940
Polonium	Po	84	1898
Potassium	K	19	1807
Praseodymium	Pr	59	1885
Promethium	Pm	61	1945
Protactinium	Pa	91	1917
Radium	Ra	88	1898
Radon	Rn	86	1900
Rhenium	Re	75	1925
Rhodium	Rh	45	1803
Roentgenium	Rg	111	1994
Rubidium	Rb	37	1861
Ruthenium	Ru	44	1844
Rutherfordium	Rf	104	1969
Samarium	Sm	62	1879
Scandium	Sc	21	1878
Seaborgium	Sg	106	1974
Selenium	Se	34	1817

Element name	Symbol	Atomic number	Date of discovery
Silicon	Si	14	1824
Silver	Ag	47	Prehistoric
Sodium	Na	11	1807
Strontium	Sr	38	1808
Sulfur	S	16	Prehistoric
Tantalum	Ta	73	1802
Technetium	Tc	43	1937
Tellurium	Te	52	1782
Terbium	Tb	65	1843
Thallium	Tl	81	1861
Thorium	Th	90	1828
Thulium	Tm	69	1879
Tin	Sn	50	Prehistoric
Titanium	Ti	22	1791
Tungsten	W	74	1783
Ununbium	Uub	112	1996
Ununhexium	Uuh	116	2000
Ununoctium	Uuo	118	
Ununpentium	Uup	115	2004
Ununquadium	Uuq	114	1998
Ununseptium	Uus	117	
Ununtrium	Uut	113	2004
Uranium	U	92	1841
Vanadium	V	23	1801
Xenon	Xe	54	1898
Ytterbium	Yb	70	1878
Yttrium	Y	39	1794
Zinc	Zn	30	Prehistoric
Zirconium	Zr	40	1789

The Most Common Elements

UNIVERSE		
Element name	**Symbol**	**Percent by mass**
Hydrogen	H	74.99%
Helium	He	23.00%
Oxygen	O	1.00%
Carbon	C	0.50%
Neon	Ne	0.13%
Iron	Fe	0.11%
Nitrogen	N	0.10%
Silicon	Si	0.07%
Magnesium	Mg	0.06%
Sulfur	S	0.05%

SUN/STAR		
Element name	**Symbol**	**Percent by mass**
Hydrogen	H	75.23%
Helium	He	23.07%
Oxygen	O	0.90%
Carbon	C	0.30%
Nitrogen	N	0.10%
Neon	Ne	0.10%
Iron	Fe	0.10%
Silicon	Si	0.09%
Magnesium	Mg	0.07%
Sulfur	S	0.04%

EARTH'S CRUST		
Element name	**Symbol**	**Percent by mass**
Oxygen	O	45.5%
Silicon	Si	27.2%
Aluminum	Al	8.3%
Iron	Fe	6.2%
Calcium	Ca	4.7%
Magnesium	Mg	2.8%
Sodium	Na	2.0%
Potassium	K	1.3%
Carbon	C	0.2%
Hydrogen	H	0.1%

OCEAN		
Element name	**Symbol**	**Percent by mass**
Oxygen	O	85.83%
Hydrogen	H	10.80%
Chlorine	Cl	1.99%
Sodium	Na	1.11%
Magnesium	Mg	0.13%
Sulfur	S	0.09%
Potassium	K	0.04%
Bromine	Br	0.01%
Carbon	C	trace
Strontium	Sr	trace
Boron	B	trace

HUMAN		
Element name	**Symbol**	**Percent by mass**
Oxygen	O	65.0%
Carbon	C	18.0%
Hydrogen	H	10.8%
Nitrogen	N	3.0%
Calcium	Ca	1.5%
Phosphorus	P	1.0%
Potassium	K	0.2%
Sulfur	S	0.2%
Chlorine	Cl	0.2%
Sodium	Na	0.1%

Trace elements include magnesium, iron, cobalt, zinc, iodine, selenium, and fluorine.

ATMOSPHERE BY VOLUME		
Element name	**Symbol**	**Percent by volume**
Permanent Gases		
Nitrogen	N_2	78.08%
Oxygen	O_2	20.95%
Argon	Ar	0.93%
Neon	Ne	trace
Helium	He	trace
Krypton	Kr	trace
Hydrogen	H_2	trace
Xenon	Xe	trace
Variable Gases		
Water vapor	H_2O	0.25%
Carbon dioxide	CO_2	0.04%
Ozone	O_3	0.01%

ATOMS AND COMPOUNDS

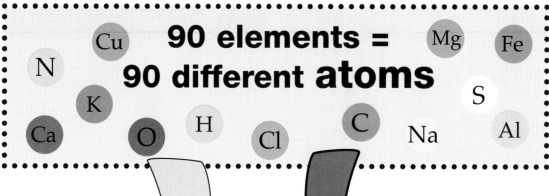

90 elements = 90 different atoms

N Cu Mg Fe K S Ca O H Cl C Na Al

Atoms combine to make new substances

1 kind of atom = elements

O O   H H

2+ kinds of atoms = compounds

O C O   Na Cl

Strong bonds between atoms = molecules

O C O

Weak bonds between atoms = ionic compounds

Na Cl

White Substances Information

Ascorbic Acid

Ascorbic acid ($C_6H_8O_6$), better known as vitamin C, is essential in the human diet. All vertebrates (animals with backbones), except primates (including humans) and guinea pigs, make their own vitamin C. Guinea pigs and primates must get it from foods, like citrus fruit, tomatoes, and liver.

Vitamin C helps body tissues grow and heal. It helps make an important protein called collagen, which is found in bones, cartilage, soft tissues, and teeth. Without vitamin C, joints hurt and grow weak, gums bleed, and teeth loosen, making it very difficult and painful to eat. The final result is death. Pretty severe consequences for not eating enough vitamin C!

The connection between these symptoms, called scurvy, and vitamin C was unknown until the late 1700s. A British naval doctor, James Lind (1716–1794), observed that eating citrus fruits cured scurvy. From that time on, sailors were required to drink lime or lemon juice to prevent scurvy.

You may have heard that a massive dose of vitamin C can prevent the common cold. According to medical studies, however, this isn't true. There are definite health benefits associated with vitamin C, but cold prevention is not one of them.

Calcium Carbonate

Calcium carbonate ($CaCO_3$) is a common mineral found in sedimentary rock, such as chalk, limestone, and marble. Carbonate rocks account for about 4% of the mass of Earth's crust. Calcium carbonate is also important in ocean ecosystems. Snails, oysters, and clams make their shells out of it. And when you see a coral reef, you are looking at the calcium carbonate skeletons of millions of tiny animals called corals.

Calcium carbonate is an inexpensive source of calcium used in the calcium pills that people take to strengthen bones. It is also used in antacids to relieve acid indigestion after a big meal. Calcium carbonate neutralizes excess hydrochloric acid found in our stomachs.

And don't forget the chalkboard. Chalk is used in classrooms around the world. Writing chalk is usually made of calcium carbonate.

Calcium Chloride

Calcium chloride ($CaCl_2$) is a salt, but not the same salt you use to flavor foods. Calcium chloride has two useful properties. When it dissolves, it releases heat. This makes calcium chloride particularly good for melting ice on roads and sidewalks. And as the salt dissolves in the melted ice, it lowers the freezing point of water so it will not freeze again. Calcium chloride has another side benefit as road salt. It does not damage plants growing at the sides of roads nearly as much as other salts.

Calcium chloride is also hygroscopic. That means it absorbs water. Because it soaks up water so efficiently, it is used to dry air and other gases. It is also spread on dirt roads. The calcium chloride absorbs water, which holds down the dust.

Calcium chloride is added to concrete to make it set up faster and cure harder. It is used as a food additive (pickles particularly) to add a salty taste without increasing the sodium content of the food. This is important for people who are on low-sodium diets and cannot eat regular salt, which is sodium chloride.

Citric Acid

Citric acid ($C_6H_8O_7$) is found naturally in almost all plants and in many animal tissues and fluids. It is important in animal metabolism. You usually think of citric acid when you think of citrus fruits like lemons, oranges, tangerines, or grapefruits. But citric acid is also found in strawberries, apples, peaches, and even brown rice, soybeans, and wheat.

Most of the citric acid used in the United States finds its way into food. It is used to preserve foods, like jams and jellies. It is also used to give texture to processed cheese. Because citric acid, like all acids, has a sharp, sour taste, it is added to candies and soft drinks to give them a sour zing. Because of its taste, citric acid is sometimes referred to as "sour salt." When you pop a sour candy in your mouth and feel your cheeks pucker and get that momentary ache in your jaw, you are having a citric acid moment.

Magnesium Sulfate

Ahhh, soak those sore feet in a nice warm bath of Epsom salts. People have soothed their sore feet in naturally occurring magnesium sulfate ($MgSO_4$) mineral springs for centuries.

Perhaps the most famous magnesium sulfate spring is located in Epsom, England. Hence the name Epsom salts. In the early 1600s, a farmer noticed that his thirsty cattle would not drink at a certain spring. He tasted the water. It was very bitter and unsuitable for drinking. But the magnesium sulfate salts were found to be relaxing and medicinal. This discovery led to the creation of the famous Epsom Spa and the first patent for medicine in England in 1698.

Today magnesium sulfate is sold in drugstores as a soaking agent for bruised, tired feet. It is also an effective laxative, providing relief from constipation. It is not too difficult to find other places where magnesium sulfate is used. It is found in fertilizers as a source of magnesium, in detergents and soaps, and in stainless steel flatware as a filler in hollow handles. It is even added to purified water to give it some taste.

Sodium Bicarbonate

Did you ever see a model volcano erupt? The "lava" was probably produced by mixing sodium bicarbonate ($NaHCO_3$) and vinegar. Sodium bicarbonate's common name is baking soda. It is added to baked goods like biscuits to make them light and fluffy. When sodium bicarbonate reacts with acid, carbon dioxide gas forms. That gas makes the foam in the model volcano and causes the biscuit dough to rise.

Sodium bicarbonate is an ingredient in many brands of toothpaste. Why is it helpful? The bacteria that grow in your mouth give off acid as a waste product. That acid eats away at the outer layer of your teeth and causes them to decay. A toothpaste with sodium bicarbonate neutralizes the acid. In fact, sodium bicarbonate is so good at fighting tooth decay that some people use it alone to brush their teeth.

Sodium bicarbonate is also good for indigestion caused by excess stomach acid. That is why it is in many over-the-counter antacids. *Antacid* literally means "against acid." Sodium bicarbonate effectively neutralizes that extra acid, forming neutral products: carbon dioxide gas, table salt, and water.

Sodium Carbonate

Sodium carbonate (Na_2CO_3), known as washing soda or soda ash, is found naturally as trona ore in only a few parts of the world. The largest deposit of trona is in Wyoming.

Hundreds of years ago, soda ash was recovered from the ash from burned seaweed. In the 1700s, sodium carbonate, recovered from seaweed ashes, was used to make glass and soap.

Today a chemical process called the Solvay process can be used to make synthetic sodium carbonate. However, this process produces a number of hazardous wastes. The preferred method of obtaining sodium carbonate is still mining and refining it from natural ore.

Sodium carbonate is important in glassmaking. To make glass, you need to melt sand (silicon dioxide). Sand melts at 1700°C. But if you add sodium carbonate to the sand, it lowers the melting temperature, making it more cost effective to produce glass.

Glass production is the largest use for sodium carbonate, but sodium carbonate has many other uses. It is used in manufacturing detergents and soaps, making paper, and treating wastewater.

Sodium Chloride

Sodium chloride (NaCl), the salt you put on food, is so important that it has played a role in the development of civilization. Salt has led to war, has served as money, and is still considered a universal symbol of hospitality. Why is salt so important? Our bodies cannot make it but must have it. Without sodium and chlorine, our bodies cannot function properly.

Sodium is essential for muscle movement, heartbeat regulation, and nerve function. Chlorine is in stomach acid, making digestion possible. It also controls the movement of water into and out of cells.

Salt is an excellent preservative, used to keep meat, fish, and vegetables from spoiling. Food that has been salted and cured can remain edible for months. Pickling foods, like cucumbers (pickles), is another method of food preservation. Food is first soaked in brine (salt water), followed by vinegar. Before refrigerators, salt was the most important preservative.

Today sodium chloride is used for much more than preserving foods. The salt industry claims that salt has more than 14,000 uses! Most of the salt used in the United States goes to make other substances, like chlorine, sodium carbonate, and hydrochloric acid.

Sucrose

Sucrose ($C_{12}H_{22}O_{11}$), best known as the refined white sugar used to sweeten foods, comes from plants. The two most important source plants are sugarcane and sugar beets.

About 70% of the sugar produced in the world is extracted from sugarcane, a very tall grass that looks like bamboo. It is grown in tropical regions around the world. The rest comes from sugar beets. Sugar beets are grown in northern, cooler climates and are roots. They look like fat, white carrots.

So why is sugar so important? It tastes good and is a source of energy in the human diet. There are 16 food calories of energy in every teaspoon of sucrose. So, if you are like most Americans, you eat 45 pounds of sugar per year! That's nine 5-pound bags, or about 2 ounces of sugar per day. That means that you eat almost 200 food calories per day of pure sucrose. That may be a bit much, but your body breaks down the sucrose into glucose that cells use as their most important fuel source. Sugar, in moderation, is an important part of our diet.

Sugar also has other uses. It is used in large quantities as a preservative in jams and jellies and is a food source for yeast in the making of bread.

FOSSweb

To find web-based information and activities related to the **Chemical Interactions Course**, go to the FOSS website: www.fossweb.com. From there, click on the "Middle School" flag, and then click the "Chemical Interactions" illustration. This will bring you to the Chemical Interactions site, which is designed for you to use at school or at home with family and friends. At this site, you will find links to related websites, lists of interesting books, a Glossary for the course, and access to the *Chemical Interactions Multimedia* program. (To access the multimedia, you will have to get a user name and password from your teacher.) For your parents, there is information about this course and the other FOSS middle school courses.

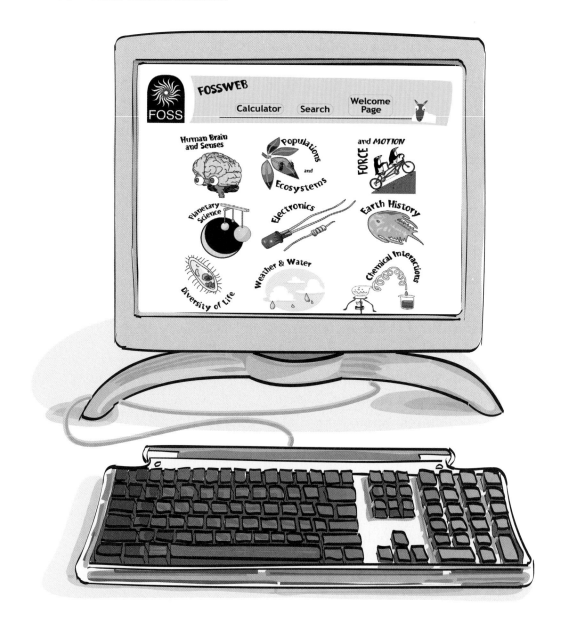

Glossary

Alchemy: The prescientific investigation of substances, including the search for ways to change common metals into gold.

Atmosphere: The layer of gases surrounding a planet.

Atom: The smallest particle of an element.

Atomic number: The number assigned to an element, based on the number of protons in the nucleus of its atom.

Average kinetic energy: Temperature.

Blood plasma: The clear, amber solution that is the liquid portion of blood.

Bond: An attractive force acting between atoms.

Calibrated: Divided into units that correspond to a standard.

Calorie: The unit of energy that will raise the temperature of 1 gram of water 1 degree Celsius.

Carbohydrate: A group of carbon-based nutrients, including sugars and starches.

Carbon dioxide gas: A compound made from carbon and oxygen; CO_2.

Chemical equation: A representation of a chemical reaction using chemical formulas.

Chemical formula: A code that represents the number and kinds of atoms in one particle of a substance.

Chemical property: A characteristic of a substance that determines how it interacts with other substances.

Chemical reaction: A process during which starting substances (reactants) change into new substances (products) with different arrangements of atoms.

Combustion: A chemical reaction, commonly called burning.

Compound: A substance whose particle is made of two or more different kinds of atoms.

Compressed: Reduced in volume as a result of applied pressure.

Concentrated: A solution with a large amount of solute dissolved in a small amount of solvent.

Concentration: The amount of solute dissolved in a measure of solvent.

Condensation: The change of phase from gas to liquid.

Conduction: The transfer of energy (heat) from one particle to another as a result of contact.

Conserved: Unchanged.

Contraction: The reduction of volume of a sample of matter as a result of cooling.

Crust: Earth's hard outer layer of solid rock.

Cyclotron: An instrument used to create new elements.

Density: The ratio of mass and volume in a sample of matter.

Deposit: The change of phase from gas directly to solid.

Dilute: A solution with a small amount of solute dissolved in a large amount of solvent.

Dissolve: To incorporate one substance uniformly into another substance at the particle level.

Dry ice: The solid phase of carbon dioxide.

Electron: A subatomic particle with a negative charge.

Element: A fundamental substance that cannot be broken into simpler substances by chemical or physical processes.

Energy transfer: The movement of energy from one location to another.

Equilibrium: A condition in which a system is experiencing no net change.

Evaporation: The change of phase from liquid to gas.

Expansion: An increase of volume.

Force: A push or a pull.

Freeze: To change phase from liquid to solid.

Fundamental: Simple and basic.

Gas: A phase of matter that has no definite shape or volume. Particles of gas fly independently through space.

Gaseous: Existing in the gas phase.

Global warming: The increase of average temperature worldwide.

Heat of fusion: Heat that causes the solid/liquid phase change without changing the temperature of the substance.

Herbicide: A plant poison.

Hydrocarbon: A group of carbon-based substances made of carbon and hydrogen only.

Insoluble: Not capable of being dissolved. Sand is insoluble in water.

Kinetic energy: Energy of motion.

Lava: Molten rock flowing on Earth's surface.

Lipid: A group of organic substances that includes oils and fats.

Liquid: A phase of matter that has definite volume but no definite shape. Particles of liquid are loosely bonded, but can flow over and around one another.

Mantle: The large rocky part of planet Earth, located between the core and the crust.

Mass: A measure of the quantity of matter.

Matter: Anything that has mass and takes up space.

Melt: To change phase from solid to liquid.

Metal: A group of elements that stretch, bend, and conduct heat and electricity.

Mixture: Two or more substances together.

Molecule: A particle made of two or more atoms that are held together with strong (covalent) bonds.

Neutron: A subatomic particle with no charge.

Nitrogen: A colorless, odorless, gaseous element that makes up about 78% of Earth's atmosphere.

Noble gas: A gaseous element that does not react with other elements.

Nucleus: The center of an atom, composed of protons and neutrons.

Octane: An eight-carbon molecule. Octane is one of the main ingredients in gasoline.

Organic compound: A large class of substances produced by organisms.

Particle: The smallest piece of a substance that is still that substance.

Periodic table of the elements: A way to organize the elements based on atomic number and chemical property.

Phase: The physical appearance of a sample of matter based on the kinetic energy of its particles. Common phases include solid, liquid, and gas.

Phloem: A plant tissue that transports nutrients to all parts of the plant.

Physical property: A characteristic of a substance that can be observed without changing it chemically, such as size, shape, density, and phase.

Potash: An impure form of potassium carbonate.

Precipitate: An insoluble product of a reaction.

Predict: To make an accurate estimation of a future event based on knowledge.

Product: A substance produced in a chemical reaction.

Protein: A group of nitrogen-containing organic substances.

Proton: A subatomic particle that has a positive charge.

Radiation: A form of energy that travels through space.

Radioactivity: Radiation given off by the elements.

Ratio: The relationship between two numbers.

Reactant: A starting substance in a chemical reaction.

Room temperature: The average kinetic energy of the particles in the air and other objects in a room.

Salt: The product that forms when a metal reacts with an acid.

Saturated: A solution with the maximum amount of dissolved solute.

Scanning tunneling microscope (STM):
An instrument that can create images of arrays of atoms.

Solid: A phase of matter that has definite volume and definite shape. The particles of a solid are tightly bonded and cannot move around.

Soluble: Capable of being dissolved. Table salt is soluble in water.

Solute: A substance that dissolves in a solvent to form a solution.

Solution: A mixture formed when one substance dissolves in another.

Solvent: A substance in which a solute dissolves to form a solution.

Sublime: To change phase from solid to gas.

Substance: A type of matter defined by a unique particle.

Transparent: Matter through which an image can be seen clearly.

Vibrating: Moving rapidly back and forth.

Volume: A defined quantity of space.

Water vapor: The gas phase of water.

Well-ordered array: A repeating pattern.

Xylem: A plant tissue that transports water and minerals to all parts of the plant.

Index

in solids, liquids, and gases, 26–27
kinetic energy, 24–25, 32–37

N

O

P

Products, 63, **105**
    in chemical reactions, 64–65
    in photosynthesis, 76–77
Propane, 75
Properties
    ancient ideas, 3
    chemical, 3–6
    phases of matter, 43–44
    physical, 3
    prediction of, 5
    of elements, 14–15
Proteins, 12, 76, **105**
Proton, 81, **105**

R

Radiation, 7, 8, **105**
Radioactivity, 7, **105**
Radium, 7
Ratio, 55, **105**
Reactants, 63, 68, **105**
    in photosynthesis, 76–77
Reaction. *See* Chemical reaction
Representing particles, 15
Road salt, 98
Rock
    lava phase change, 44–45
    properties of, 16–17
Room temperature, **105**
    at equilibrium, 36
    particles in motion, 24
Rutherford, Ernest, 80

S

Safety practices, 89
Salinity, 58
Salt, **105**
    calcium chloride, 98
    dissolving in water, 51
    element in the ocean, 11, 52, 58, 59
    sodium chloride, 100
Saturated solutions, 57, **105**
Scanning tunneling microscope (STM), 80, **105**
Scurvy, 97
Seaborg, Glenn, 81

Seawater, 11, 52, 58, 59. *See also* Ocean
Silicon
    atoms, 80
    in the ocean, 59
    on Earth, 10
6-mercaptopurine (6-MP), 84, 85
6-MP. *See* 6-mercaptopurine
Sky, elements in, 11. *See also* Atmosphere
Soda ash, 100
Sodium
    discovery of, 7
    in humans, 12, 100
    in the ocean, 11
    particle, 15
Sodium bicarbonate, 15, 99
Sodium carbonate, 100
Sodium chloride, 100
    in the ocean, 11, 52
    particle, 15
Solar system. *See* Universe
Solid, **106**
    expansion and contraction of, 28–31
    heat and —, 27, 45–46
    freeze/melt temperature, 46
    particles in, 18
    phase change, 42–43, 44–45, 46–48
    phase of matter, 16–22, 43–44
Soluble, 53, **106**
Solute, 51–53, **106**
    concentration, 54–58
Solution, 51, **106**
    concentrated solutions, 54–55
    cough drops in water, 49–50
    defined, 53
    dilute, 55–56
    for life, 52–53
    mixture and —, 51
    on Earth, 51–52
    saturated solutions, 57
    transparent, 51, **106**
Solvay process, 100
Solvent, 51–53, **106**
    concentration and, 54–57
Star. *See* Sun
Starches, 76
Stardust, 9